Robert M. Koerner and Joseph P. Welsh
CONSTRUCTION AND GEOTECHNICAL ENGINEERING
USING SYNTHETIC FABRICS

J. Patrick Powers
CONSTRUCTION DEWATERING: A GUIDE TO THEORY AND
PRACTICE

Harold J. Rosen
CONSTRUCTION SPECIFICATIONS WRITING:
PRINCIPLES AND PROCEDURES, Second Edition

Walter Podolny, Jr. and Jean M. Müller
CONSTRUCTION AND DESIGN OF PRESTRESSED
CONCRETE SEGMENTAL BRIDGES

Ben C. Gerwick, Jr. and John C. Woolery
CONSTRUCTION AND ENGINEERING MARKETING
FOR MAJOR PROJECT SERVICES

James E. Clyde
CONSTRUCTION INSPECTION: A FIELD GUIDE TO PRACTICE,
Second Edition

Julian R. Panek and John Philip Cook
CONSTRUCTION SEALANTS AND ADHESIVES, Second Edition

Courtland A. Collier and Don A. Halperin
CONSTRUCTION FUNDING: WHERE THE MONEY COMES
FROM, Second Edition

James B. Fullman
CONSTRUCTION SAFETY, SECURITY, AND LOSS PREVENTION

Harold J. Rosen
CONSTRUCTION MATERIALS FOR ARCHITECTURE

William B. Kays
CONSTRUCTION OF LININGS FOR RESERVOIRS, TANKS,
AND POLLUTION CONTROL FACILITIES, Second Edition

Walter Podolny and John B. Scalzi
CONSTRUCTION OF CABLE-STAYED BRIDGES, Second Edition

Edward J. Monahan
CONSTRUCTION OF AND ON COMPACTED FILLS

*Construction Of and On
Compacted Fills*

Construction Of and On Compacted Fills

Edward J. Monahan
Bloomfield, New Jersey

A Wiley-Interscience Publication

John Wiley & Sons

New York • Chichester • Brisbane • Toronto • Singapore

Library of Congress Cataloging in Publication Data:

Monahan, Edward J.
    Construction of and on compacted fills.

    (Wiley series of practical construction guides,
0271-6011)
    "Wiley-Interscience publication."
    Bibliography: p.
    Includes index.
    1. Building.   2. Fills (Earthwork)   3. Soil
stabilization.   4. Foundations.   I. Title.   II. Series.
TH153.M56 1986   624.1′5   85-29644
ISBN 0-471-87463-9

Printed in the United States of America

10 9 8 7 6 5 4 3 2 1

To

My wife, Mary-Jean, for her constant support in general, and especially for supporting me with regard to important career decisions coinciding with the period of the writing of this book.

and to

David McKay Greer, my early "teacher" in the professional practice of geotechnical engineering, with the hope that he will get away from drilled piers for a while, so that he can complete *Reminiscences of an Arkansas Traveler*.

A foolish consistency is the hobgoblin of little minds.

—RALPH WALDO EMERSON

A respect for complexity is the beginning of wisdom.

—DEAN RUSK

Everything should be made as simple as possible, but not simpler.

—ALBERT EINSTEIN

Series Preface

The construction industry in the United States and other advanced nations continues to grow at a phenomenal rate. In the United States alone construction in the near future will approach two hundred billion dollars a year. With the population explosion and continued demand for new building of all kinds, the need will be for more professional practitioners.

In the past, before science and technology seriously affected the concepts, approaches, methods, and financing of structures, most practitioners developed their know-how by direct experience in the field. Now that the construction industry has become more complex there is a clear need for a more professional approach to new tools for learning and practice.

This series is intended to provide the construction practitioner with up-to-date guides which cover theory, design, and practice to help him approach his problems with more confidence. These books should be useful to all people working in construction: engineers, architects, specification experts, materials and equipment manufacturers, project superintendents, and all who contribute to the construction or engineering firm's success.

Although these books will offer a fuller explanation of the practical problems which face the construction industry, they will also serve the professional educator and student.

M. D. MORRIS, P.E.

Preface

This book is in the Wiley Series of Practical Construction Guides. It deals
with compacted fills and is written principally to be helpful to the construction contractor. In order to serve that purpose, however, it is necessary to
provide practical and useful information to *all* nonspecialists who traditionally are involved with or affect the planning, plans, specifications, and
execution of earthwork construction. Such nonspecialists include the architect, the structural engineer, and the fill inspector, whose activities have
a strong impact on the contractor's performance and profits.

Recent spectacular failures of man-made earthen structures have been
extremely costly in terms of both dollars and lives. On June 5, 1976, the
305-ft Teton Dam in Idaho collapsed. Seventeen months later, on Sunday,
November 6, 1977, at approximately 1:30 A.M., the Kelly Barnes Dam near
Taccoa, Georgia, failed. The resulting flood took 39 lives. As a direct result
of the Taccoa Dam disaster, a national dam inspection program was initiated.

The principal cause of both failures has been judged by geotechnical
experts to have been piping, a progressive interior deterioration of the
earth structure associated with seepage forces of sufficient magnitude to
dislodge and slowly transmit particles along with flowing water.

As imposing as the losses associated with these nationally reported dis-
asters are, I am convinced that these losses pale in comparison to damage
caused by the countless thousands of defective earth structures—or fills—
that are constructed annually. The number of cracked sidewalks, pavements,
floor slabs, walls and columns, and basement walls must be prodigious.
Improperly backfilled trenches resulting in traffic hazards and damaged
utilities can be both costly and dangerous. The costs escalate significantly
when litigation becomes a factor, especially where personal injury or deaths
are involved.

A peculiarity of earthwork construction is that many nonspecialists who
are routinely involved frequently do not consult with geotechnical specialists
for advice regarding filling operations. I believe the reasons for this are
(1) a belief that filling operations are simple and straightforward and do

not require the attention (and expense) of specialists and (2) a preoccupation with the (eventual) visible aspects of the structure such as architectural considerations and structural design.

The principal reasons for this undesirable situation are (1) the neglect of the topics of soil compaction and fill control in *required* undergraduate courses in civil engineering and civil engineering technology and (2) that little information seems to be available to *nonspecialists* involved in earthwork aspects of engineered construction.

I hope that this book will correct or at least alleviate the foregoing problems.

EDWARD J. MONAHAN

Bloomfield, New Jersey
March 1986

Acknowledgments

I would like to acknowledge the contributions of many former students at the Newark College of Engineering to the work in this book. Because of the short time of my association with each of them in the academic setting, it was generally not possible to bring to a satisfactory conclusion most of the projects that were initiated. But it was the constant stimulation they provided that was responsible for my broad study of compacted fill technology, represented especially by the content of Chapter 5. Many of these former students are now in professional practice "simulating field conditions" instead of getting sick of hearing the phrase; my best to them.

I also cite the valued contributions of Harvey Feldman, Chief of Laboratory; Tippetts, Abbett, McCarthy, and Stratton, New York, for his thorough and professional review of the manuscript, executed in a very timely and helpful fashion. Certain additions and modifications were made as a direct result of his comments, and the book was, I think, markedly improved as a result. The chapter glossaries, for example, were added because of his perception that too many terms were left unexplained in the manuscript copy he reviewed, particularly since the book was directed (mostly) to nonspecialists. He also suggested a short description of dynamic compaction.

Thanks to Tom Bellatty, early professional colleague, former MVAP (most valuable adjunct professor), and erstwhile square root student, for his insights regarding the role of "observers" in the implementation and enforcement of specifications. Ed Dauenheimer's comments and help in obtaining information with respect to "professional liability loss control" (Section 9.2) are also appreciated.

The assistance of Dan Morris, consulting editor for the Wiley Series of Practical Construction Guides, was invaluable, but most of all his attitude with respect to allowing "me to be me" (within reason) is most appreciated.

Finally, I am indebted to Mrs. Pearl Gordon, who typed the manuscript with much skill and patience and with great good judgment. She was always "there" when various deadlines were imposed along the way.

E. J. M.

Contents

1. Introduction, 1

   1.1   *Purpose and Scope, 1*
   1.2   *Importance and Nature of Earthwork, 2*
   1.3   *Important Definitions, 3*
   1.4   *The Role of Index and Engineering Properties, 8*
   1.5   *Glossary, 11*

2. Avoiding Costly Blunders, 13

   2.1   *The Practical Value of Knowledge of the Historical Development of Soil Compaction, 13*
   2.2   *Early Empirical Approaches, 13*
   2.3   *Rational Approach, 14*
   2.4   *Standard Proctor Density, 14*
   2.5   *Modified Proctor Density, 15*
   2.6   *Load-Bearing Fills Building Codes, 16*
   2.7   *Summary, 16*

3. Basics of Soil Compaction Curves: Laboratory Procedures, 18

   3.1   *Compaction Defined, 18*
   3.2   *Spectrum of Soil Types, 20*
   3.3   *Curve Locations and Shapes and Their Practical Meaning: Moisture and Energy Effects, 22*
   3.4   *ASTM Compaction Requirements, 23*
   3.5   *Summary, 26*
   3.6   *Glossary, 26*

4. Major Problems in Compacted Fill Technology: Proposed Solutions, 28

   4.1   *Standard–Modified Ignorance, 28*
   4.2   *The 95 Percent Fixation, 31*

4.3 *Changing Borrow, 39*
4.4 *Problems Evolving from Traditional Practice, 42*
4.5 *Cost and Time Pressures: A Summary, 44*
4.6 *Glossary, 45*

5. Applied Research and Development, 46

5.1 *Effects of Specific Index Property Variations, 46*
5.2 *Waste Materials as Fills, 52*
5.3 *Artificial Fills, 53*
5.4 *Effects of Mechanical Laboratory Compactors, 66*
5.5 *Density Gradients, 67*
5.6 *Geostick Correlations, 67*
5.7 *Percent Compaction Specifications for Clay Fills, 68*
5.8 *Summary, 69*
5.9 *Glossary, 69*

6. Fills and Fill Compaction, 71

6.1 *Strength, Stability, and Imperviousness:*
      Contrasting Requirements, 71
6.2 *Potential Problems with Earth Structures, 72*
6.3 *Controlled and Uncontrolled Fills, 86*
6.4 *Nonstandard and Special Fills, 86*
6.5 *Compactors and Lift Thicknesses, 88*
6.6 *Energy and Moisture Control, 96*
6.7 *Glossary, 97*

7. Compaction Specifications, 99

7.1 *Typical Specifications, 100*
7.2 *Implementation and Enforcement, 111*
7.3 *Nontechnical Aspects of Specifications, 114*
7.4 *Specification and Project Evaluation, 115*
7.5 *Glossary, 118*

8. Fill Control Procedures—Inspection, 119

8.1 *Field-Density Testing, 119*
8.2 *The Compleat Field Man, 141*
8.3 *Case Histories, 151*
8.4 *Glossary, 175*

9. Techonomics, 176

    9.1 Engineering Design, 176
    9.2 Legal Costs, 177

10. Update at Press Time, 181

    10.1 Weight-Credit Update, 181
    10.2 Very Large Jobs, 184
    10.3 Annotated References for Self-Study, 186
    10.4 Additional Seminars, 190
    References, 190

Index, 193

*Construction Of and On
Compacted Fills*

1

Introduction

1.1 Purpose and Scope

The purpose of this book is to present helpful information on soil compaction and fill control to nonspecialists. The information should also be useful to academic colleagues, and especially so if I am successful in convincing them of the importance of a greater emphasis in their courses, especially those which are heavily attended by nonspecialists. Because of the focus noted, established on the basis of 26 years of observations in both the classroom and in consulting practice, I assume no prior knowledge of soil compaction, or for that matter, soil mechanics. A careful study of Chapters 1, 2, and 3 is recommended for a better appreciation and understanding of that which follows.

Chapter 4 presents the major problems in fill control, and solutions are proposed for each problem. Some are strictly technical in nature, while others deal with much more sensitive matters ranging from professional ethics to human relations. Observations made about the latter are bound to be controversial. However, I am convinced that what is said needs to be said, openly and forthrightly, in order to have any chance of correcting certain practices detrimental to good engineered construction.

In order to provide credibility and validation to the assertions made regarding major problems, case studies are used. In all cases, names, locales, and other possibly embarrassing details are omitted.

Chapter 5 describes some of my unpublished research results and suggests further research efforts. I believe investigations can be pursued at many levels—funded research, undergraduate and graduate laboratory courses, and undergraduate and graduate projects and theses. I hope that geotechnical consultants adopt some of the suggestions and improve and extend certain procedures to augment their recommendations for, and supervision of, filling operations. In some cases, for example, the suggestions regarding a compaction data book (for dealing with the problems of *changing borrow*), their ability to accumulate large amounts of data makes it feasible to develop, refine, and validate methods rapidly.

Chapter 6, "Fills and Fill Compaction," deals in detail with the technological aspects of all types of compacted fills, and includes a section on potential problems with earth structures, with an emphasis on earthen dams. The intent is to provide information helpful to those involved with the National Dam Inspection Program, thus serving the dual purpose of improving capabilities in the inspection and remediation of completed earth structures and in the full range of activities needed for design, inspection, and construction of new earthworks. Chapter 7 complements Chapter 6 with a description of all details relating to compaction specifications.

Chapter 8 (plus some aspects of Chapter 7, notably Specification Evaluation, Section 7.4) is intended to serve as a manual for fill control procedures: it is written expressly for the typical inexperienced young geotechnical engineer or engineering technologist.

The book closes with two short chapters: "Techonomics" and "Update at Press Time." The former is a coined term relating to the important link between technical and economic decisions in design and construction. The last chapter includes some material that evolved, was discovered, or just "seemed to be sensible additions" during the latter stages of completion of the manuscript.

As a departure from typical format, chapter glossaries are provided at the end of each chapter, to explain certain terms used in the chapter, which it was judged could not be explained fully within the text without adversely affecting readability. Terms are listed in order of appearance in the chapter. I suggest that the reader review each chapter's glossary before reading the chapter of interest.

1.2   Importance and Nature of Earthwork

The importance of soil compaction and fill control can be emphasized by one simple declaration: that almost no significant engineered construction occurs without the movement of soil from one place to another. Furthermore, it should be the nature of good engineered construction that parties become involved in earthwork operations in the following sequence: the geotechnical engineer, the architect and structural engineer, the fill inspector, and the construction contractor. A most important nonspecialist is, of course, the owner or client. Unfortunately, as noted in the preface, the geotechnical engineer is too often left out of the sequence. All too often, he is called in (late) to correct a bad situation or to act as an expert witness when it is already too late for correction.

In the ideal sequence, the geotechnical engineer explores and evaluates the subsurface conditions through a logical process of analyzing soil (or rock) *index properties,* and then, as judgment dictates, determines appropriate *engineering properties* through laboratory and/or field testing. With a knowledge of the loading conditions that are to be imposed by the proposed structure, the engineer prepares recommendations for foundation type, methods of foundation construction, and allowable bearing capacities at particular foundation levels. Such recommendations are broadly based upon determining safe loading intensities on foundation elements (e.g., footings) of various sizes that will not result in either a *bearing capacity failure* or *unacceptable settlements.* More often than not, the recommendations will entail excavation and filling operations, requiring recommendations for *quality* (texture) and *condition* (compacted density) of the fill. Assuming compliance with specifications, the fill is then judged to have a certain allowable bearing capacity.

The structural engineer is one of the primary recipients of the geotechnical engineer's report. In addition to the obvious responsibility of structural design, it is his typical responsibility to prepare working drawings, including foundation drawings. The preparation of construction specifications becomes a natural extension of this work, typically as notes on the drawings and (for large projects) separate, additional documents. In the usual situation, the structural engineer will need to consult architectural drawings and geotechnical reports in the conduct of his work and communicate with the architect and geotechnical engineer when questions develop requiring their attention and expertise.

Construction engineers and contractors are, of course, the "doers." They bring to fruition the studies, concepts, designs, drawings, and specifications of the architect, the geotechnical engineer, and the structural engineer, thus completing the sequence of operations of engineered construction.

1.3 Important Definitions

Soil Compaction (Field). The reduction of void spaces (densification) of lifts of fill by the direct application of load, impact, and/or vibration, usually with a suitable type of compaction equipment. Lift thicknesses vary from several inches (for clays and silts) to perhaps 2 ft for free-draining fills (sands and gravels). [Note: One should not confuse soil compaction with consolidation, the long-term reduction of void ratio of a natural soil, usually saturated, thick deposits of soft clays or silts beneath the water table. This is usually accomplished by the application of static surface loading (called surcharges), resulting in the slow drainage of porewater from the subsurface stratum. Thus, one principal difference is that compaction is direct and immediate.]

Soil Compaction (Laboratory). The compaction of a small but representative soil sample, obtained from the field, in a steel mold of standard

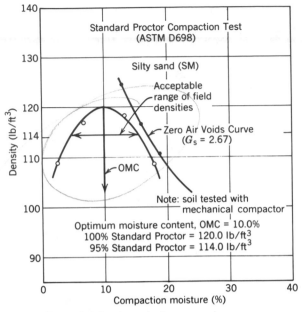

FIGURE 1.1   A typical compaction curve.

size. The soil is compacted usually in layers, commonly by dropping a hammer of specified weight through a specified distance a specified number of times. The energy of such compaction is chosen to simulate that of field compaction with rollers. The moisture content is varied for a series of filled molds, thus generating a *compaction curve* for the soil. A typical curve is shown in Figure 1.1.

Granular, Cohesionless Soils.   Gravels, sands, and "clean" silts (those possessing no plasticity).

Cohesive, Plastic Soils.   Clays, clay-silt mixtures, organic soils.

Drainage Quality Designations.

1. *Free-draining* soils—gravels, coarser sands, and mixtures thereof.
2. *Marginally draining* soils—finer sands, clean silts, and mixtures thereof.
3. *Impervious* soils—clays, clay–silt mixtures. (Note that no soil is impervious, so the term is used in a relative rather than absolute sense.)

Gemicoss.   James Kilpatrick, one of my favorite columnists and lingophiles, suggests that if there is a legitimate need for a new word, invent one. Herewith, *gemicoss.*

It is defined as any soil that is a combination of soil types in such proportions as to raise questions pertaining to contrasting engineering properties, for example, cohesionless vis-à-vis cohesive, plastic vis-à-vis nonplastic, free-draining vis-à-vis marginally draining. It may also be helpful to tell you how I invented the word: As a child, I learned that all (most?) English words required one or more vowels. Thus G is for gravel, C is for clay, SS is for sand and silt. E, O, and I, of course, are needed vowels. I might also represent inorganics, the soil minerals. In the Unified Soil Classification System, O is used to represent organics, and M to indicate silts. EM also stands for Ed Monahan.

There are many examples illustrating the need for the concept of the gemicoss. Often a chart or a formula has a principal limitation of being applicable to a certain soil type, that is, sand or clay. Blow counts, for example, are used for determining allowable bearing capacities *for sands*. For clays, one often uses the unconfined compressive strength. But what does one "do" with a sand–clay mixture? Or a sand–silt–clay mixture? There is no simple answer, other than to say that good engineering judgment must be employed.

Essentially Granular (or Cohesive) Soils.   Texture that predominates in the context of dictating the overall behavior of the soil. This terminology is a natural extension of the notion of a gemicoss. After carefully evaluating the texture (sizes *and* plasticity) of the soil, one may be forced to decide whether it is essentially granular or essentially cohesive in order to justify the use of a particular design chart or formula.

For example, if a soil is composed of 60% highly plastic clay (by weight), with 40% sand and gravel, the soil would be an ECS, an essentially cohesive soil, since the sand and gravel particles are, for all practical purposes, merely isolated or "suspended" in a clay matrix, and their presence is essentially irrelevant to the overall engineering behavior of the soil. Note that this example does not fit the classification of a gemicoss, because the conclusion (ECS) is fairly evident. A gemicoss might be 40% medium plastic clay, 60% sand and gravel. Such a mixture raises questions pertaining to contrasting engineering properties.

As a contrasting example, if a soil is 75% gravel and sand, with 25% clay of high plasticity, it would be rated an essentially granular soil (EGS), inasmuch as it is reasonable to assume that the preponderance of granular particles are in contact. The significant percentage of highly plastic clay would act as "binder."

Thus, in summary, the simple auxilliary classification system that I propose includes five categories: granular, cohesionless soils; essentially granular soils; essentially cohesive soils; cohesive (plastic) soils; and the gemicoss.

Fill.   The soil that is selected for use at the site. It may be from a site excavation or "imported" from elsewhere. In the latter case, the fill is termed *borrow*.

Fill Quality Indices.   Texture: grain sizes and grain size distribution of cohesionless soils, or plasticity (usually plasticity index, or PI) of cohesive soils.

Fill Condition.   Density (pounds per cubic foot, pcf) or relative density (percent), a measure of potential settlement. For clay soils, potential *expansion* is often an important additional condition to consider.

Permanent Stability.   A concept referring to the fundamental question of whether the fill (or any soil) will remain stable under present *and future* conditions that may reasonably be expected to be imposed during the economic life of the structure. Thus, a very dense, free-draining soil, which is to be permanently contained (laterally), is the most "permanently stable" soil, for it cannot settle significantly, it is not susceptible to seepage pressures, nor will it expand when wetted. Conversely, a cohesive (clay) soil can settle unacceptably if not compacted sufficiently but can expand detrimentally upon (future) wetting if overcompacted. The pressures associated with such expansion can be substantial, causing damage to highway pavements and even heavy structures. Clays and silts are also susceptible to disturbance by seepage pressures (piping, uplift) because of their relative imperviousness. Silts are also susceptible to frost action.

Simulation of Field Conditions.   This is an important concept concerning the rationale that should be used for all decisions relating to laboratory testing. Thus, what one does to the soil sample in the laboratory should, in all practical respects, simulate what will be done to the soil in the field. As one of the more obvious examples, the method and energy of compaction in the laboratory compaction test should approximate that of construction rollers used in the field.

Index Properties.   Indicates the general nature of the subsurface problems that must be confronted. Such index properties are (or can be) obtained routinely and at modest expense in a typical subsurface investigation. Figure 1.2 shows, in simple flowchart form, how index properties should be used. Note that large jobs, requiring the expenditure of larger amounts of money, would justify the expenditure of larger amounts for field or laboratory testing to determine engineering properties needed for design. Engineering properties are defined as those measuring stress, strain, and strain rate on representative soil samples that enable computation of specific numerical values of bearing capacity and settlement. Thus, path ABC provides a design that is more "reliable" than path AC, but at considerably more cost. Path AC, for small jobs, where extensive testing is not economically justified, utilizes index properties directly for design purposes, usually by entering a chart to select allowable bearing capacities, rather than the more expensive ABC route. Of considerable practical importance, moreover, is the fact that

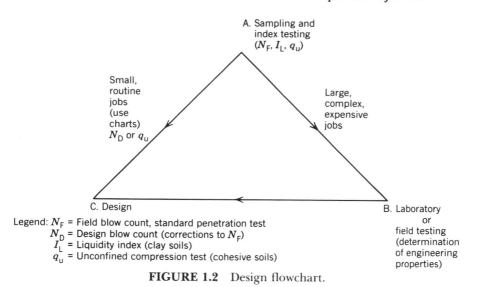

FIGURE 1.2　Design flowchart.

index properties, if obtained and properly evaluated, serve as the basis for a rational testing program (step B). In summary, index properties serve a twofold purpose and are indispensable tools in geotechnical engineering.

Blow Count, Density, and Relative Density.　Illustrative of the foregoing description of index and engineering properties are the definitions of blow count, density, and relative density, particularly as related to compaction and fill control and the stability and bearing capacity of granular, or essentially granular, soils.

When a driller is sampling a soil that is expected to be an essentially granular soil, a sampling device is driven into the soil and a blow count, N (blows per foot) is obtained. Clearly, the blow count (with any appropriate corrections) is an indicator of density; the higher the blow count, the higher the density.

Relative density, a laboratory-determined *engineering property*, is a percentage of the practically obtainable maximum density for the particular soil. Thus, a soil of 100% relative density would have a negligible potential for future settlement. Furthermore, given the thickness of the soil, a knowledge of the relative density would allow for the computation of a specific, numerical settlement estimate. A judgment could then be made as to whether the computed settlement would be tolerable.

In this example, the blow count is an indicator (or index property) signifying the *general* condition of the soil (e.g., very loose, loose, medium-dense, dense, very dense), whereas the relative density (involving the added expense of laboratory testing) is an engineering property.

With respect to fills, knowledge of the relative density and the compacted density—which the designer can control by appropriate specifications—permits similar computations for settlement.

Rather than list and describe all index and engineering properties in this introductory chapter, terms will be defined where first used or will be included in the chapter glossaries.

1.4 The Role of Index and Engineering Properties

If the texture and blow counts of a soil being considered as potential borrow are determined from exploratory drilling and sampling operations, what practical information can be derived from such index data?

Case I

Case I involves a well-graded gravelly sand, with approximately 25% fines of significant plasticity. [PI on minus-40 fraction is determined to be 30. Liquid limit (LL) is 65.] Blow count range, 35–60.

Definitions and Discussion.   Fines are all soils passing the No. 200 sieve, thus silts or clays. If fines are judged by simple and quick field tests to be significantly plastic (clayey rather than silty), Atterberg Limits tests are ordered on that portion of the soil passing the No. 40 sieve: the minus-40 fraction. Result is PI = 30; LL = 65. Such fines would be rated "high plasticity", CH in the Unified Soil Classification System. In summary, this soil is a dense to very dense well-graded essentially granular soil, but containing cohesive clay binder. In a real situation, the locale of the potential borrow area would, of course, be known. Thus, soil maps, geologic maps, and perhaps personal knowledge of the area (geologic, topographic, and land-use, for example) would be available to augment the data from the exploratory program. The soil described is typical of a glacial till, soil deposited directly by a glacier in a mechanical fashion as opposed to alluvial deposition (by say, glacial meltwater). This accounts for the wide range of sizes, as contrasted to uniform sizes associated with the sorting action of flowing water. Thus, one could reasonably expect to encounter boulders in the area. Confirmation could be obtained by reconnaissance, particularly by inspection of road cuts in the vicinity.

Before proceeding with a listing of the practical and potentially very valuable information which can be gleaned from the foregoing, I feel it is necessary to add some commentary here regarding a very important subject: soil (and rock) descriptions as they pertain to formal classification systems. There are several such systems. The most commonly used, in the sense of broadest acceptance in the United States, are the Unified system and the

AASHTO system (the American Association of State Highway and Transportation Officials). The best system, in my opinion, is the Burmister system, developed by Professor Donald Burmister of Columbia University. It is outside the scope of this book to present details concerning classification systems, or to editorialize extensively about the merits or shortcomings of each. Suffice it to say the the Burmister system, used extensively by many of my colleagues in professional practice in the New York/New Jersey area, forces the person using the system to be precise in estimating and then describing the texture of a soil sample, but not unreasonably or unnecessarily so. With concentration and practice, one can rather quickly develop the skill of estimating the proportions of constituents (gravel, sand, silt, clay) to within about 15% accuracy, the gradation of each granular constituent (coarse, medium, fine), and the approximate plasticity of any fines (silts, clays) present. Using only visual and tactile senses (sight, feel, smell—even taste, with one soil scientist I know!) formal, unambiguous descriptions can be written in accordance with the explicit rules governing the system.

The description given for Case I, although it will serve its purpose here, is not in accordance with any formal system. Thus, in professional practice, it would be of little value, since "well-graded," "essentially granular," and "containing clay binder" have no well-defined meaning to other potential users. A soil classification system, then, is seen as a universally agreed-upon language, wherein all words and symbols used have specific definitions. When people describe soils without adopting a recognized system, no one, save themselves, can be sure of what they mean. Unfortunately, it is all too common to see soil descriptions in practice, generally written by one of our nonspecialists, which are ambiguous, vague, and often virtually useless because of usage of terminology not belonging to any recognized system.

For a brief yet excellent treatment of the Unified and AASHTO systems (and some others) refer to Ritter-Paquette (1960), Chapter 6. For a complete description of the Burmister system, see Burmister (1958) and related publications.

With definitions, discussion, and commentary regarding the importance of careful soil classification, the following list briefly gives some representative examples of valuable information pertaining to Case I.

1. The potential borrow will be expensive to excavate because of its density and texture. Bulldozers or power shovels would probably be required. Scrapers would require "assistance" in excavating the soil.

2. Boulders (if confirmed) could create significant additional cost, depending upon frequency and size; better check road cuts.

3. The soil does not drain well (D_{10} size* in clay range). Hence, rain on borrow area of significant amounts would probably create delays and work

---

*D_{10} size, also called Hazen's effective size, can be used to estimate soil permeability.

stoppages, also possible trafficking problems. Check topographic maps for surface drainage patterns and access roads.

4. The soil has excellent texture from the standpoint of potential bearing capacity, depending of course on degree (energy) of compaction, but will have poor drainage characteristics.

5. There is slight-to-medium potential frost action.

6. If compacted sufficiently, postconstruction settlements are no problem, but watch out for overcompaction, which could produce expansive potential because of highly plastic fines.

7. Best compactor is probably a rubber-tired roller.

Case II

Case II involves a sand with coarser sizes predominant, fine gravel present and no significant silty materials evident. (Sieve test showed fines fraction to be 7%; dilatancy test on fines confirm no plasticity—NP.) Blow count range, 15–25.

Definitions and Discussion.   On the basis of preliminary inspection and evaluation, this soil appeared to "qualify" as a *select fill*. In New Jersey, one of the principal textural qualifications for a select fill is that it contain not more than 12% passing the No. 200 sieve. The sieve test was ordered to determine (particularly) the percent fines. The 7% confirms the select fill classification.

Geologically, this select fill is probably of alluvial origin. Judging by its low blow counts, it is a recent alluvium (mapped as AR on soil maps published by SCS, the Soil Conservation Service, a division of the U. S. Department of Agriculture) and has the following characteristics.

1. Excellent texture. It is becoming scarcer (except, of course, in shoreline areas) and is thus a valuable "commodity"; it should therefore be reserved for special purposes and used sparingly. Typical special purposes are trench bottoms to support pipelines and graded filters in earth-dam construction.

2. Free-draining. Thus, not susceptible to seepage pressures, frost action. If properly compacted, it is permanently stable (except Richter 8's).

3. Easily excavated (low blow counts, no cohesion).

4. No laboratory compaction test is necessary (or even sensible). Write compaction specifications on basis of percent relative density. Compact by puddling.

5. Best compactor would be a heavy vibratory roller.*

*I will explain these comments more fully in subsequent chapters.

1.5 Glossary

Unconfined compressive strength.   The compressive stress at failure on a cylindrical specimen, usually expressed in tons per square foot (tsf); commonly performed on sampled, natural soils of a cohesive nature (e.g., clays). There is no reason, however, that such a test cannot be performed on compacted samples of suitable cylindrical dimension.

Plasticity (plasticity index, cohesive).   Property of material (soil) that enables it to be deformed rapidly without volume change, without rupture, and without elastic rebound. The shape, of course, does change; a sculptor's modeling clay is a familiar example. The specific range of moisture contents over which the soil acts in this way is called the plasticity index (PI). All soils exhibiting a measureable PI are cohesive, meaning simply that the particles "stick together" naturally by other than (or in addition to) capillary forces. Clays, clay mixtures (clay with other soil types), and organic soils are cohesive.

Granular (cohesionless).   Coarser particles of soil (gravel, sand, "clean" silts) that exhibit no plasticity or cohesiveness.

Piping.   The dislodging and movement of finer soil particles under the influence of seepage of water through the soil, starting usually as a slow process, and leading eventually to greater distress and, possibly, failure (explained more fully in Section 6.2.1).

Frost action (frost heave).   The seasonal heaving (winter) and recession (spring breakup) of frost susceptible soils, notably silty soils. A serious problem in highway and other route design. (See also Section 6.2.1.)

Alluvial.   A geologic term referring usually to soils transported or deposited by water (streams, glacial meltwater, etc.).

D_{10} Size.   This means that 10% of the soil (by weight) is finer than that size. Thus, it may be thought of as an indicator of the fineness of a soil; an important index in assessing the general seepage quality of a soil. (See also Section 5.1.2.)

Dilatancy test.   A simple, diagnostic test of a soil to determine degree of plasticity. The finer fraction of the soil in question, generally that which passes the No. 40 sieve, is mixed with enough water to mold it into a soft, saturated ball about ½ in. in diameter. The ball is formed into a wafer shape and placed in the palm of the hand. The heel of the hand is then gently jarred repeatedly with the other hand, while one observes the surface changes (if any) of the wafer-shaped soil. A ready and quick shiny, glassy change in appearance (i.e., within seconds) constitutes a positive dilatancy test, and means that the soil is nonplastic (n.p.). If one squeezes the hand with the pat, the surface will immediately become dull in appearance; resumed tapping will quickly re-create the shiny appearance. (The squeezing shears the soil and allows the surface fluid to quickly drain away, thus assuming the dull appearance. The tapping consolidates the soil, and the water film reappears on the surface to yield the shiny appearance.) The quickly positive result identifies the soil as a "clean" silt (as opposed to, say, a clayey silt) or a fine sand/silt mixture. The longer it takes to produce any reaction (called dilatancy), the higher is the degree of plasticity. This and other field tests are used by experienced soils people to assess the plasticity rating of the soil in a qualitative way. For confirmation, the PI is determined by laboratory analysis for the liquid limit and the plastic limit, both of which are moisture contents. The numerical difference is the PI. (For details concerning field testing or laboratory testing, see References No. 3 and 20, respectively.)

Graded filter.   A soil filter that provides a gradual (graded) transition for water flow, to prevent piping, for example. (See also Section 6.1.)

Richter 8.   The Richter Scale is used to rate the intensity of earthquakes. Data obtained from seismographs located throughout the world results in the rating, which is on a logarithmic scale (base 10); thus a Richter 6 is 10 times more intense than a Richter 5, and so forth. A Richter 8, used somewhat facetiously in the text, would be a rare and very major earthquake of devastating effect.

2

Avoiding Costly Blunders

Before describing details of soil compaction and fill control, it will be helpful to focus briefly on a few broad topics, knowledge of which will help to avoid major, potentially costly mistakes.

2.1   The Practical Value of Knowledge of the Historical Development of Soil Compaction

Ordinarily, expositions on historical development of a subject are presented in textbooks rather than practical guides. However, as I hope I will be able to demonstrate, the blunders that occur today in earthwork construction are strongly rooted in the past. While I am certain that Santayana was referring to much more esoteric subjects than soil compaction and fill control, his admonition that "those who cannot remember the past are condemned to repeat it" applies well.

Key periods include the 1920s and the early 1950s, which may be regarded as significant turning points in the development of the state of the art.

2.2   Early Empirical Approaches

Prior to the 1920s, indeed into antiquity, earthwork construction was strictly empirical. It is a matter of record, for example, that the Scot, John L. McAdam, stabilized soils by driving herds of sheep over areas of soft subgrades prior to construction of embankment and pavement sections (circa 1800). The modern sheepsfoot roller derives its name from this early construction practice. One can cite the Appian Way as a classic example of ancient road building of remarkable quality and stability, clearly accomplished without benefit of modern soil mechanics theory and practice.

Soil mechanics was started by Karl Terzaghi in the 1920s, with the decision of this great engineer–scientist to develop a scientific approach to treating soil as an engineering material. A contemporary of Terzaghi's, R. R. Proctor, is regarded by most geotechnical engineers as the originator of principles dealing with a rational approach to soil compaction, with his publication "Fundamental Principles of Soil Compaction" in *Engineering News Record,* August–September, 1933.

2.3 Rational Approach

Among Proctor's major contributions was a recognition of the need for a laboratory compaction test to control filling operations. Observing that soil moisture and compaction energy were two of the most important variables affecting compaction, emphasis was placed upon a determination of these relationships. Proctor's first decision was most probably the choice of a laboratory compaction energy that would simulate field energies available at the time, that is, 1920-vintage rollers. The method of applying the energy in the laboratory was through the use of a falling hammer, most likely chosen for computational simplicity rather than simulation of the kneading action of field rollers. (As is frequently the case, the exact simulation of field conditions in the laboratory is not practical or feasible, but it is essential to good testing practice to strive constantly toward this end, and to use good judgment in recognizing divergencies and their possible effects on results.) To simulate the field practice of compacting soils in lifts, Proctor specified that the laboratory test would simulate this factor by compacting in layers; thus laboratory layers simulate field lifts. Representative soil samples would be compacted in layers in a mold, at varying moisture contents, at energy chosen to simulate that of the field rollers.

2.4 Standard Proctor Density

The most pertinent details of Proctor's original compaction test are shown in Table 2.1. By conducting this test at varying moisture contents, the laboratory curve for most soils (exception: free-draining soils) took the form shown in Figure 1.1. The form of the curve suggested the terminology— optimum moisture content—meaning moisture content that would produce 100% Proctor density, or more generally, moisture content that would result in most efficient compaction in the field. Thus, if the natural field moisture happened to coincide with this value—not a likely event—field compaction efficiency would be ideal.

The engineers of the 1920s, undoubtedly recognizing the impracticality

TABLE 2.1. The Proctor Compaction Test, 1920s

Mold Size	Layers	Hammer	Hammer Fall	Blows/Layer	Total Energy E_c
$\frac{1}{30}$ ft^{3a}	3	5.5 lb	12 in.	25	12,400 ft-lb/ft^3

^{a}(4.6 in. × 4.0 in. diameter)

of specifying field compaction at precisely optimum moisture content, adopted a range of acceptable density values, as shown in the figure. Someone decided, apparently on a reasonable but totally arbitrary basis, that 95% Proctor would be acceptable. Thus the phrase "shall be compacted to 95% Proctor" was born. The contractor would have a reasonable range of densities (and field moistures) within which he could perform satisfactorily and efficiently. In practice, soils that were too wet could be dried by scarification, and soils that were too dry could be wetted by sprinkling, whereupon efficient compaction could be accomplished.

Field density tests of the rolled fill could be done to determine compliance with specifications.

2.5 Modified Proctor Density

In the late 1940s and 1950s, work progressed on the greatest road-building project in history: the 40,000-mile Interstate Highway System. By that time, it was possible to compact soils in the field to much higher densities needed to support the heavier vehicles that would use the system. Accordingly, the personnel of the American Association of State Highway and Transportation Officials (AASHTO) developed a laboratory compaction test of higher energy to simulate field energies of the heavier rollers. Details are shown in Table 2.2. Note that the "new" test was simply a modification of Proctor's original test, hence the two names for the same test: modified AASHTO and modified Proctor. To further differentiate, the original Proctor was renamed "standard Proctor," and various organizations provided number designations. These are summarized in Table 2.3.

The names and designations for the tests should be used with precision, since the modified Proctor energy is about 4 ½ times that of standard Proctor.

TABLE 2.2. The Modified Proctor (AASHTO) Compaction Test, 1950s

Mold Size	Layers	Hammer	Hammer Fall	Blows/Layer	Total Energy E_c
$\frac{1}{30}$ ft^3	5	10 lb	18 in.	25	56,000 ft-lb/ft^3

TABLE 2.3.   **Compaction Test Names**

Names	Energy (ft-lb/ft^3)
Standard Proctor (or standard AASHTO) ASTM D698 AASHTO T99 British Standard 1377: 1948	12,400
Modified Proctor (or modified AASHTO) ASTM D1557 Modified AASHTO[a]	56,000

[a]AASHTO is the modern name for AASHO, reflecting greater emphasis on transportation and transportation planning, notably mass transit in urban areas, upon essential completion of construction of the Interstate Highway System.

2.6   Load-Bearing Fills/Building Codes

To further illustrate and emphasize the major difference in compaction energy of the two tests, you should be aware that standard Proctor densities have been historically regarded as *nonload-bearing,* at least in the sense of supporting structural loads such as footings. Building codes written before about 1947, for example, typically required all footings to bear on n.g.— natural ground. Standard Proctor densities represented the practical limits of construction capability at the time. Soils compacted to standard Proctor densities have some supporting capability, such as for parking areas and lightly traveled secondary roads, but they were never intended for support of heavier structural loads.

Today, it is common practice to specify footings to bear on compacted fill, typically compacted to 95% modified Proctor densities. According to Sowers (1979), modified Proctor energy is "comparable to that obtained with the heaviest rollers under favorable working conditions." I have performed field density tests on soils compacted to modified Proctor energies. I call them "blister densities," meaning that a person not used to such activity will develop blisters in the course of digging several holes with a garden spade. While this is not a very scientific terminology, the writer had found that this designation sticks most firmly in the mind of a young engineer assigned to the task.

2.7   Summary

To avoid blunders, remember:

1. The major distinctions between *standard* and *modified:*
   (a)   Standard energy: 12,400 ft-lb/ft^3.
          Modified energy: 56,000 ft-lb/ft^3.

(b) Standard is *nonload bearing* (light loads acceptable).
Modified is *load bearing.*

2. Use or interpret compaction test names (or number) designations carefully and precisely. Get angry (as I do) when someone says, "95% compaction."

3

Basics of Soil
Compaction Curves:
Laboratory Procedures

As a result of Proctor's work, and subsequent developments in the state of the art by others, it is necessary to have a thorough knowledge of soil compaction theory and laboratory procedures in order to plan and execute construction of compacted fills in the field. Before proceeding to the field, then, we must look to the books and laboratories.

3.1 Compaction Defined

When a typical borrow is compacted in the laboratory, an interesting phenomenon is observed at very low moisture contents (0–5%, approximately): the curve dips (see Figure 3.1). The moisture at this low level forms thin capillary films that develop tensile stresses upon impact of the compaction hammer. The tensile stresses cause a corresponding intergranular compression, resulting in a sudden increase in the effective strength of the soil. Figure 3.2 illustrates the situation schematically. As moldfuls of representative borrow are compacted at higher moisture levels (2, 3, 4), the capillary films thicken and become an effective lubricant. At even higher moisture levels, however, a point of diminishing returns is reached, where the addition of more water results in decreasing densities (5, 6, 7). This behavior suggests what I believe to be the most illuminating definition of effective soil compaction: *the expulsion of air from the soil mass*. This definition allows for the easy understanding of an observation that is vital to good qualitative fill control in the field: *that soil undulation in front of and behind*

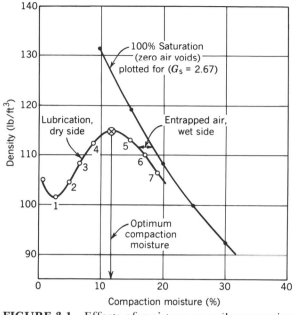

FIGURE 3.1 Effects of moisture on soil compaction.

a field roller means the soil is too wet for efficient compaction at the energy level (roller weight) being used. Figure 3.3 shows the analogy between the corresponding laboratory and field behavior. When the soil is being compacted efficiently (lubrication or "dry" side, Figure 3.1), air is expelled from the soil upon impact of the hammer *in quantities larger than the volume of water added.* However, soil has an "air permeability" such that at some high level of moisture (less air voids), the moisture cannot escape under impact of the hammer. Instead, the entrapped air is energized and lifts the soil in the region around the hammer. There are characteristic sounds associated with the laboratory compaction, a solid "thud–whoosh" (escaping air) for efficient compaction, and a not-so-solid "galumph" when the soil is too wet for compaction. In the field, the roller energizes the entrapped air in the soil, causing the undulation corresponding to the lifting of the soil upon

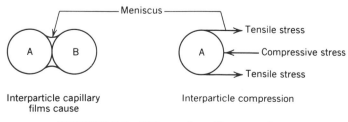

FIGURE 3.2 Effects of capillary tension.

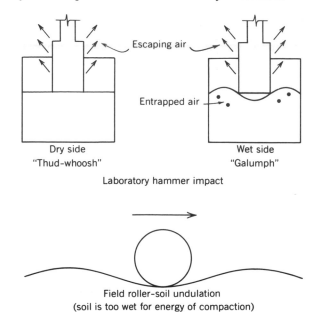

Escaping air

Entrapped air

Dry side
"Thud-whoosh"

Wet side
"Galumph"

Laboratory hammer impact

Field roller-soil undulation
(soil is too wet for energy of compaction)

FIGURE 3.3   Effects of excess moisture: laboratory and field correlation.

hammer impact in the laboratory test. Efficient compaction in the field is, of course, characterized by an absence of undulation. The most helpful way of viewing what has happened when a soil is too wet is to recognize that water has been added without any expulsion of air, resulting in an increase in void space (i.e., a decrease in density).

3.2   Spectrum of Soil Types

Figure 3.1 shows entrapped air as the distance between the wet side of the curve and the line of 100% saturation (or zero air voids). The expression for dry density as a function of percent saturation (S) and compaction moisture content w_c is

$$\gamma_d = \frac{G_s\,\gamma_w}{1 + \dfrac{w_c\,G_s}{S}}$$

where $\gamma_d =$ the dry density of the compacted soil,

$G_s =$ the specific gravity of soil minerals,

$\gamma_w =$ the unit weight of water,

$$w_c = \text{compaction moisture content, and}$$

$$S = \text{percent saturation}$$

For cohesionless soils (quartzitic), $G_s = 2.65$, and for most common clay minerals, $G_s = 2.70$. With $S = 1$, a zero air voids curve can be plotted which, with little error, would be representative of all soils. Furthermore, the zero air voids curve is an envelope under which the compaction curves for all soils, compacted at any energy level, will fall, with wet sides approximately parallel to the saturation line envelope. An array of such curves is shown in Figure 3.4. Curves for five soil types, intended to represent the spectrum of all soils commonly used in earthwork, are shown, each compacted to the two common energies: standard Proctor and modified Proctor. Curves 1 and 5 represent TALB: typical allowable load-bearing borrow. Texturally, it would ordinarily be well-graded and predominantly granular, but with significant percentage of fines, preferably of sufficient plasticity to impart overall cohesive qualities (i.e., "clay binder"). In my suggested auxiliary classification system described in Chapter 2, it would be an EGS, and would be marginally draining or impervious, depending principally on the exact quantity and plasticity of fines. Note that such a soil can be compacted to densities approaching 140 pcf with the modified Proctor energy: "blister densities."

Curves 2 and 3 depict free-draining sands, or select fills. Curves 6 and 9, and 7 and 10, represent "lean" clays and "fat" clays, respectively. "Lean" and "fat" are commonly used by soils engineers to designate relative degrees of plasticity. Thus, a fat clay is highly plastic and, for our purposes here, we can say it has a greater affinity for water. I call clay mineral particles

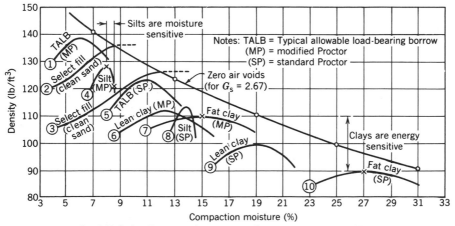

FIGURE 3.4 Compaction curves for spectrum of soil types.

of this type "thirsty little devils." Lean clays have low to medium plasticity, with PI's less than 40 and liquid limits less than 50%. Casagrande's Plasticity Chart is the most common method of classifying plastic soils (Lambe, 1951, p. 27).

Finally, curves 4 and 8 represent silts, fines that are nonplastic (n.p.) or, at most, slightly plastic (PI 0–5).

In summary, as strongly inferred by the preceding definitions, it is important to know not only the amount (percentage) of fines in a soil mixture, but also the plasticity of those fines. Construction problems and costs can be markedly different for different textures. The engineering behavior of the compacted soil also varies greatly for different textures.

3.3   Curve Locations and Shapes and Their Practical Meaning:
Moisture and Energy Effects

Of all the illustrations in this book, I regard Figure 3.4 as the most important; important enough to be studied over and over until its details are fixed in memory, including approximate numerical values. Following are explanations pertaining to details.

3.3.1   Variations in Optimum Moisture Contents

The optimum moisture content of a soil is inversely related to compactive energy. [There is a widely held misconception that a soil has an "inherent," unchanging optimum moisture content. This misconception probably developed because, for the early years (1920–1950), only one energy was used, standard Proctor.] *A soil can have almost any optimum moisture content, depending upon the compactive energy used.*

Although the range of compaction moistures shown in Figure 3.4 extends from 3 to 33%, observe that the range of optimum moisture contents for TALB has a considerably narrower range, about 5–12%. Moreover, if the borrow is to be load-bearing (as defined earlier) it should be compacted to energies approximating modified Proctor, narrowing the range to say, 5–9%. Thus, with knowledge of the foregoing, one could estimate the approximate optimum rather closely. Indeed, if one commits the set of curves shown to memory, even approximately, *a fairly close estimate could be made for the optimum moisture content for any soil compacted at any common energy level* (i.e., standard or modified Proctor).

In the foregoing commentary, note that the term TALB refers initially to the textural quality of the fill. The other criterion for load-bearing capability is, of course, the *condition* of the fill—its compacted density. The literature frequently refers to controlled fills and uncontrolled fills without further distinction, but it is helpful and frequently of considerable practical (cost) consideration to be aware that a fill can be controlled in two ways:

quality (texture) and condition (placement density). This matter is discussed more fully in subsequent chapters.

3.3.2 Moisture and Energy Sensitivity

The shapes of curves have practical significance. Silts and clays are two extremes, with silts steep and clays relatively flat.* As indicated on the figure, *silts are moisture-sensitive,* meaning simply that a small change in field moisture will effect a major change in compacted dry densities achievable *at the same compaction energy.* (Note that two field options for correction would be possible: changing the soil moisture or changing the compaction energy.) *Clays are energy-sensitive,* meaning that relatively small changes in compaction energy can produce large changes in density (curves 7 and 10). Thus, if one area of a site is inadvertently overcompacted by channelization of construction traffic, larger future differential settlements will result. Stated another way, it is probably better to have the entire fill compacted to (say) 92% modified Proctor than half 92 and half 98. In fact, serious overcompaction of clays can produce a fill that will expand on future wetting, thus subjecting any structure on the site (or the fill itself, as with earth dams) to intolerable deformations.

3.3.3 Free-Draining Soils: Select Fills

Free-draining sands, or select fills, have no discrete optimum moisture content (curves 2 and 3), for the simple reasons that they *are* free-draining. Thus, if one attempts to generate a compaction curve of the more typical peaked, symmetrical shape, the added moisture, at higher levels, would simply drain or be ejected from the mold upon impact of the hammer. In the field, therefore, fills must be compacted either completely dry or "puddled." Except in very arid areas, the former condition is impractical, so puddling is normally used, a process of compacting the soil by applying an excess of water, usually with a hose.

Comprehensive details regarding field compaction of all soil types will be presented in Chapter 6, "Fills and Fill Compaction."

3.4 ASTM Compaction Requirements

To conclude this section, preparatory to a detailed presentation of the nature of the problems in earthwork construction, it will be helpful to describe

---

*Throughout this broad discussion of the entire spectrum of soil types, the terms "silt" and "clay" refer to soils that are essentially silt (or silty), clay (or clayey), as defined earlier. A more specific treatment of percentages of various soil constituents and their effects is presented in Chapter 5.

briefly the steps generally required to develop laboratory compaction curves needed for field compaction control. These steps generally follow those required by ASTM—the American Society for Testing and Materials.

1. Obtain a large bag sample of the soil at the site or borrow area that is judged to be representative. Generally, 100 lb or more is required in order to prepare several batches at different moisture contents. Recompaction of the same batch, by extruding a compacted soil, pulverizing it, and adjusting to a different moisture level, is not allowed. Research has shown that significant errors can occur if the recompaction process is used. The error is probably more pronounced at higher compaction energy (modified Proctor), since the higher energy would undoubtedly cause more "chipping" of larger soil particles upon repetitive recompaction. Angular soils are probably more susceptible to such chipping. Modification of the soil in this way during recompaction in effect produces a different soil (grain size distribution) upon each recompaction. Since the soil in the field will not be so modified, the notion of simulating field conditions is violated without justification. The only apparent justification is the ease of handling and time saved recompacting a small (10 lb) representative sample.

Successive recompaction of cohesive soils should also be avoided because an unrepresentative rearrangement of soil particles would result, creating a more pronounced parallel layering than would be the case for soil compacted in the field. This layering is called a "dispersed" structure. Its effects are described more fully in Section 6.2.2.

2. Air-dry the soil overnight, spreading the soil out to a depth of 3–4 in. to facilitate drying.

3. Separate the soil into approximately equal, representative batches, by a quartering or splitting process.

4. Determine the weight and moisture content of each batch.

5. Determine the dry weight of soil retained on the No. 4 sieve, and record as percent of batch. The maximum particle size used in the compaction mold is logically based upon mold size. If a very large percent of "No. 4" is present, it is sensible to use a larger mold and a correspondingly larger exclusion size. A 6 in. diameter mold, with a $+\frac{3}{4}$ in. exclusion, is common. If this option is exercised, the energy should be increased to about 55 blows per layer for the modified Proctor test. To account for the material excluded in the compaction test, a correction should be made to field density values, based upon differences in excluded material between the laboratory and field (density hole) samples.

6. Add moisture to each air-dried batch to produce moisture contents in 2–3% increments in the anticipated range encompassing optimum moisture content. Considerable experience is necessary to accomplish the foregoing in an efficient manner, ideally producing two points on either side of optimum. However, consideration of Figure 3.4 and related explanations

should enable one to estimate the approximate moisture contents required. To effect the actual soil preparation at the desired moisture contents, an instant moisture balance, used with the following formula, is very helpful.

$$W_A = \left(\frac{W_d - W_k}{1 + W_k}\right) W_{TB}$$

where W_A = weight of water that must be added to a soil of known moisture content to raise it to some higher (desired) moisture content,

w_d = desired moisture content,

w_k = known moisture content (as determined by the "instant" moisture balance*), and

W_{TB} = total weight of wet batch of soil (i.e., soil + water).

The instant moisture balance is a table-top device that enables one to determine the moisture content of a soil in several minutes. In the typical case, one would determine the air-dried moisture content, w_k (known moisture content), and then plug into the formula successively the values of moisture content desired, w_d, to determine W_A values necessary to generate the curve. Because of the need for thorough mixing of each batch, which will diminish the moisture content by aeration, it is necessary to overshoot the desired moisture content. Two to 3% overshoot is recommended. Thus, if one wishes to produce a thoroughly mixed soil of moisture content $w_d = 9\%$, use $w_d = 11.5\%$ in computing the amount of water to be added. These procedures are also very helpful for filling any "holes" in the compaction curve that are discovered after plotting the curve. Alternative procedures may introduce unnecessary errors, or are time-consuming and frustrating (trial-and-error, the usual method).

7. Place the wetted, prepared batches in a large can with a lid or in sealed plastic bags. Cure overnight in a humid room. (This allows the moisture to soak into the soil particles, thus providing closer simulation of field conditions.)

8. Compact each batch in accordance with the required energy (e.g., Table 2.1 or Table 2.2). Determine weight of each compacted soil sample and, from the extruded cylinder of soil, obtain representative chunks for determination of compaction moisture, w_c. Weigh chunks and place in oven (105°C) overnight, or until dry weight does not change.

*Cenco Model No. 41101.

Compute the dry density:

$$\gamma_d = \frac{(W/V)}{1 + w_c}$$

where W = weight of soil in mold,

V = volume of mold, and

w_c = compaction moisture (expressed as decimal).

9. Plot the compaction curve and zero-air-voids curve (see Figure 3.1).

3.5 Summary

1. Undulating fill means that the fill is too wet for the energy of the compactor being used.

2. Optimum moisture content of a soil decreases with increasing compaction energy.

3. Figure 3.4 is an excellent aid in estimating the approximate range of optimum moisture content for any soil.

4. Silts are moisture-sensitive (steep curve).

5. Clays are energy-sensitive.

6. Select fills have no discrete optimum moisture content.

7. ASTM laboratory compaction requirements are extensive and time consuming. Counting acquisition of the sample and delivery time, one day each for air drying, curing wetted samples, and moisture content determinations, a minimum of four days would be required. With other common delays, such as those caused by heavy laboratory testing demands, a full work week is often required to obtain the results of one test. This feature of compaction testing can become a matter of considerable importance, and relates to one of the major problems in compacted fill technology, that of changing borrow.

3.6 Glossary

Compaction moisture content, w_c. The moisture content w of a soil is the ratio of the weight of water in the soil to the weight of the solids (minerals), expressed in calculations as a decimal, but often referred to otherwise as $n\%$. The compaction moisture content, then, is simply the moisture content at which the soil is compacted.

Percent saturation, S. The ratio of the volume of water in a soil to the volume of voids, expressed as a decimal in calculations.

Quartering. A technique used for reducing a large sample of soil to a representative sample of a desired smaller size, done by spreading the soil to about a 3-in. deep pile, and cutting it into quarters, then one quarter into quarters, etc.

Splitting.   Another technique for obtaining representative samples of desired size. In this case a "soil splitter," an apparatus designed for the purpose (available from various suppliers of laboratory equipment), is used.

4

Major Problems in
Compacted Fill Technology:
Proposed Solutions

Fill control is the art of ensuring the proper selection, placement, and compaction of soil to provide permanent stability for the structure it is to support. This control is usually provided by preparing written specifications governing the texture and compacted density of the fill. End-result specifications normally call for some percentage of the laboratory-determined maximum of some recognized standard, for example, 95% standard Proctor or modified Proctor. Methods specifications deal with matters such as required field compaction moisture, lift thicknesses, and compaction equipment. Almost all specifications incorporate features of both types requiring a *target value density* and specifying means of attaining that value. Of logical necessity, however, some flexibility should be incorporated in methods specifications to accommodate day to day changes in the field (notably weather, and thus moisture); methods of attaining the target value will require adjustment.

Following are descriptions of problems in fill control, their causes, and some suggested remedies. Some brief case histories are included to emphasize some of the major problems, problems that all too frequently involve major blunders resulting in increased construction costs, construction failures, or unpleasant confrontations leading to professional embarassment and, in one case, unethical behavior.

4.1 Standard–Modified Ignorance

Because of the relatively recent transition from standard Proctor to modified Proctor energy, as described earlier, there exist many old specifications in the job files of many engineering firms calling for Proctor densities (now called standard Proctor). These, of course, were written either for roadway embankments, parking areas, or for some purpose other than support of structural foundations, since building codes did not generally permit such practice. Also, because of the very newness of soil mechanics itself, there are many older engineers who did not take formal courses during their college years and, for various reasons—mostly an understandable preoccupation with their own specialty interests—have not become aware of the very significant difference between the two terms, standard and modified.

Of such a combination of circumstances are major blunders made.

4.1.1   Case Study 1

In the early 1960s, I was assigned to a fill control job involving the placement of about 150,000 yards of fill.* The assignment was made on something of an emergency basis. The soils consulting firm for which I worked received a phone call on Friday, requesting the service of a fill-control specialist for the following Monday morning. I was given the address of the firm and the name of the engineer requesting the service. No other information about the job was provided.

Within a few days of arrival, the following situation and information was discovered:

1. Stripping and grubbing had been completed (removal of trees and brush).
2. Site preparation involved a cut–fill operation to convert a gently sloping site into a flat area, with the fill constituting approximately 60% of the site.
3. Filling operations had already started. One compactor was available, a small steel-wheeled roller of obvious early vintage, including old-fashioned spoked wheels and a gravity-feed gas tank.
4. No borings were available.
5. No investigation had been made to locate borrow. In fact, it appeared that the problem had not been considered.
6. Written specifications, amounting to about two-thirds of a typed page, were supplied. Ninety-five percent standard Proctor density was the specified target value for the fill.
7. The thickness of the fill varied from zero (at the no-cut, no-fill line) to approximately 14 ft at the downhill region of the site.

As you may infer, the situation described was to be fraught with difficulties. However, in order not to lose focus on the purpose of this section, major problems of fill control, descriptions of routine problems and operations are deferred to subsequent chapters.

*A common slang term for cubic yards.

About two weeks into the job, an engineer (*not* a soils specialist) representing the ultimate site tenant visited the site, principally to observe site preparation activities. During an early conversation, a comment was made about 3000 lb/ft^2 floor loads.

It became evident that something was seriously wrong: 3000 psf floor loads on a 14-ft fill were not even remotely compatible with 95% *standard* Proctor energy. The local consultant's engineer suggested that the necessary higher densities could be obtained by getting the contractor to use heavier rollers and more passes—that, if this were done cleverly, the contractor could be made to produce *modified* energy densities without being apprised of any specification changes.

Telephone consultation with my project engineer led quickly to an agreement to ignore the unethical proposal, and a job meeting was arranged for the purpose of addressing the problem and its ramifications. The specifications were rewritten with the major change to modified Proctor energies, and the contractor was properly compensated for the required extra work.

As might be expected, there were no conversations among the parties concerned regarding the unethical oral proposal described. Thus, the circumstances leading to this blunder must—with one exception — be surmised. It is certain that the engineer to whom the writer reported, and who supplied the "original" specifications, knew very little about soil compaction. The specifications were probably obtained or derived from the files. Finally, it is speculated that the retention of the firm to supply consulting services and inspection for soils work dealing with site preparations was a last-minute decision based upon a late-developing awareness of the need for such services.

As may be inferred from these circumstances, the earthwork contractor's superintendent was also largely ignorant of the geotechnical engineering aspects of soil compaction and fill control. As surprising as this may seem, in view of its overriding importance to successful bidding, it has been my experience that such ignorance is common. Even big earthwork contractors sometimes lack even a rudimentary knowledge of the major differences between standard and modified Proctor, notably differences in cost and effort. Many small contractors cannot achieve modified Proctor densities commonly called for in specifications for load-bearing fills, because they do not own, nor can they afford, the heavy rollers needed to produce such densities efficiently. (See Case Study 2, following.)

The solution to the problem of standard–modified ignorance is to disseminate knowledge widely among persons who are not soils specialists and who are involved in earthwork. This can be done effectively in two ways: (1) through publication of concise, practical information in a practice-oriented medium that is explicitly advertised as "useful in practice" and is widely marketed,* and (2) by convincing academicians of the importance

---

*The phrase "useful in practice" is an officially recognized keyword in the Geodex Information Retrieval System. It is, however, unlikely that very many persons who are not soils specialists have access to this system or, indeed, are even aware of its existence.

of greater emphasis on the subject in courses that are composed of a large majority of students specializing in areas other than soils (i.e., structures, construction, transportation, and "other"). This class composition would always be the case in the undergraduate *required* course in soil mechanics offered at accredited engineering colleges. Moreover, records compiled over a period of seven years in a graduate course, "Shallow Foundations," at the New Jersey Institute of Technology (NJIT) indicate 77% were nonspecialists (86 of 111). Data of recent years show an even higher percentage: over three years, 84% were nonspecialists (62 of 74).

4.2 The 95 Percent Fixation

Since the reasonable but otherwise arbitrary choice of 95% Proctor densities as the target value for fill control work of the 1920s was made, there has developed a fixation on the part of specification writers to require this percentage compaction, irrespective of loadings, fill thickness, or other factors that should logically influence compaction requirements.

There are a number of causes for this practice: a reluctance to specify anything different than is "customary," nonspecialists writing compaction specifications, and the fact that there is no widely accepted rational method for specifying percentage compaction appropriate for specific conditions.

4.2.1 Case Study 2

A small project involved the expansion of a structure for light industrial use. The proposed expansion was to extend to a sloped area as shown in Figure 4.1. I recommended orally that a continuous wall footing be founded in compacted fill (level A), thus avoiding the apparent greater cost of a couple of lifts of concrete block (for footing at level B). The contractor agreed.

I performed a geotechnical site investigation and prepared a report that was distributed to the structural engineer and the contractor. The pertinent excerpt from my report is: "It is recommended that this fill be placed in lifts 8 in. to 12 in. thick and compacted to *93% modified AASHO*."

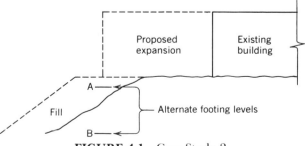

FIGURE 4.1 Case Study 2.

In time, after receiving my report, the structural engineer prepared the structural drawings, among which was included the foundation plan containing the following:

Foundation Notes

1. The contractor shall read the soils report dated . . . by Edward J. Monahan, P.E. and shall familiarize himself with the site and the conditions as outlined in the report. . . .

Compacted Fill Notes. . .

4. Fill shall be placed at its optimum moisture content in uniform layers not more than 8 in. thick after compaction, and each layer shall be thoroughly compacted to a density not less than *95 percent* of the density prescribed in ASTM D1557-66 T. [emphasis mine]

5. The compacted fill shall be placed under the supervision of a licensed professional engineer . . . who will submit a certification of the fill.

Commentary

Because of the light loadings and relatively thin fill required, I recommended less than the usual percent compaction, that is, 93 rather than 95. Note, however, that the structural engineer, while acknowledging the writer's geotechnical expertise (Foundation Note No. 1), could not accept the "unusual" lower compaction percentage recommended.

Also note that the specifications called for a "certification of the fill" by "a licensed professional engineer" and "placement supervision" (Compacted Fill Note No. 5). As it developed, the contractor ignored the supervision requirement. The fill was placed and compacted, a continuous footing was poured, and several lifts of block were constructed, all without the specified supervision. With some misgivings and dismay, and with certainty about the outcome, I performed a field density test on the compacted fill in the area immediately adjacent to the wall. As expected, the fill was nowhere near the specified 95% modified AASHO density. In fact, the field density test indicated approximately 67% modified AASHO. As a result, I had no recourse but to insist upon the costly procedure of ripping out the wall and the fill, and rebuilding the wall foundation in accordance with the specifications for placement and certification.

In addition to the principal illustration of the "95% fixation" problem, this simple case history presents evidence of the other problems: (1) that many contractors do not generally appreciate or understand what the stringent requirement of modified energies (at high percentage levels) entails in terms of required machinery and placement methods (moisture, lift, thickness, number of required passes), (2) that nonspecialists prepare unrealistic soil specifications—unrealistic with respect to compatibility with load requirements and unrealistic with respect to demands on contractors. In addition to the unnecessarily high energy requirement, the phrase "*at its* optimum moisture content in uniform layers not more than 8 in. thick *after* compaction [emphasis mine]" contains questionable verbiage. What does "*at its* optimum moisture content" mean? Taken literally, *any* divergence from optimum is not permissible. Since the important end-result requirement is 95% modified AASHO density, some latitude in methods should be permitted. Indeed, it is logical that completely rigid methods specifications make it impossible to comply with an end-result specification!

The phrase *its* (optimum moisture content) makes one suspect that the writer of the specifications believes that the soil has *an* optimum moisture content—a typical misconception of nonspecialists. (See earlier commentary, Chapter 3, "Basics of Soil Compaction Curves.")

The phrase "8 in. thick *after* compaction", again taken literally, would require the fill inspector to reject the fill if the compacted surface elevation revealed said "violation," even if the soil density target value was satisfactory. Lift thickness requirements should therefore specify an acceptable range of *loose* placement thickness.

Finally, one wonders what the word "thoroughly" means in Compacted Fill Note No. 4.

These latter criticisms may seem niggling compared to the more serious error of over- and underspecifying compaction energies, but collectively such careless phrasing in specifications can create unnecessary friction in field operations.

4.2.2   Case Study 3

A large construction project for a college medical complex included structures and parking areas with a wide variety of loading and site preparation requirements. The resident soils engineer, an evening graduate student in my course, "Shallow Foundations," raised a question in class that had developed on the job that day. Work had concluded in one area of the site that involved compacting soil to 95% modified Proctor density, and attention was shifted to another area of the site requiring fill. In this case, however, apparently because of lighter loadings and other factors allowing for less stringent fill compaction, the written specifications called for 95% *standard* Proctor density. The problem was that no compaction test had been done for other than modified Proctor energy, so no target value could be determined, that is, no standard Proctor curve was available. The question: what to do? The recourse of the student (resident engineer) was to call his office for advice, which was to use 93% modified. The engineer asked his supervisor, why 93%? The answer: "Because I said so."

It is evident that 95% standard Proctor is equivalent to some lesser percentage of modified, but no clear-cut, rational way appeared to exist to determine the specific lower percentage for a given soil; in effect, an experienced person simply uses his own judgment. Thus, in Case Histories 2 and 3, the judgment was, in each case, "use 93."

One means of correlating standard and modified percentages is simply to compile data on a variety of soils that have been compacted at both energy levels. Thirty-three soils were thus evaluated by undergraduate students at NJIT (Mardekian, Rowbotham, and Facente, 1972). The percentage modified corresponding to 95% standard ranged from a low value of 81.4 for a "clay" to 92.2 for a "Florida sand," with an average of 86.6. Only 4 of the 33 samples exceeded 90%, each of which was an essentially granular soil, predominantly sand. Accurate textural descriptions and related index properties were not obtained for most of the soils, but it appears safe to conclude that, in general, 95% standard Proctor conforms to percentages considerably less than the 93 value used in Case Histories 2 and 3. Since the level was chosen in both cases strictly on the basis of experience, the figure of 93 was "sensibly conservative."

As a result of my concern with the problems described, I developed an explicit method for specifying percentage soil compaction (Monahan, 1974). Because of research and additional analyses, certain improvements have been made. Following is a description of the method, with explanations of the need for further modifications, and suggestions for research to improve and extend the method toward broader applicability in practice.

4.2.3   A Method for Specifying Percentage Soil Compaction

Almost all site preparation work involves the compaction of soil. And many specifications for such work require 95% Modified AASHTO densities, irrespective of the intended use of the fill. In cases such as lightly loaded warehouses, parking areas, and subgrades and embankments for secondary roads, such a stringent compaction requirement is unnecessary and consequently not economical. In special cases, where heavy loads and/or structures highly sensitive to differential settlements are involved, it may be advisable to specify a percentage compaction in excess of the "the usual" 95%.

In order to provide a means for determining economical compaction requirements, the following procedures have been developed.

Figure 4.2 is the familiar design chart recommended by Terzaghi and Peck for footings on sand; allowable bearing pressures indicated are based upon the settlement criterion (which governs for all but the loosest sands and/or very small footings). The usual use of the chart is by abscissa-entry, that is, one enters the chart with a first-estimate footing size, and reads the allowable bearing capacity corresponding to the *in situ* blow count N for the natural soil deposit as determined by the Standard Penetration Test. (Note: Some corrections to the blow count may be necessary.) The essential difference between this procedure and one where compaction is to be performed is that control can be exercised over the eventual bearing capacity of a controlled fill. Thus, one can reverse the procedure and ask,

What "blow count" is needed to provide a specified (or desired) bearing capacity, and what percentage compaction is needed to develop a density which would yield such a blow count?

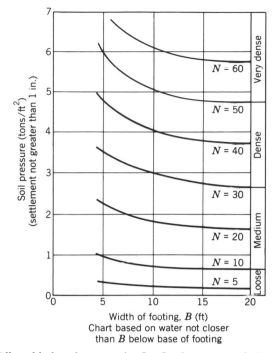

FIGURE 4.2   Allowable bearing capacity for footings on sand. From Peck, Hanson, and Thornburn, Foundation Engineering, 1st ed., Copyright John Wiley & Sons, 1953.

Figure 4.3 suggests a method for answering such a question. The graph represents the results of a compaction test for a TALB, the most common of which is modified AASHTO (ASTM 1557). Added to the plot is a scale of void ratios, with limiting values of e_L and e_D corresponding to relative densities of 0 and 100%, respectively (loosest and densest). The relative density is given by

$$D_R = \frac{e_L - e_n}{e_L - e_D}$$

where e_n is usually thought of as the natural *(in situ)* void ratio. However, in this procedure, the value is interpreted as the required void ratio of the compacted fill e_c:

$$D_R = \frac{e_L - e_c}{e_L - e_D}$$

It remains to define procedures for estimating e_L and e_D. For very large, important jobs, involving either large cost or protection of human life (as

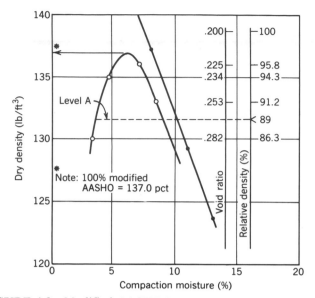

FIGURE 4.3   Modified AASHTO compaction test for a TALB.

with earth dams or high retaining walls) ASTM procedures for e_L and e_D determinations are recommended. However, as a practical matter, it is thought that quicker, approximate, and much less expensive methods are justified for most routine applications, Indeed, it has been my goal to develop a design method that will combine validity with acceptability to users. Other criteria are simplicity, quickness, and low cost. Finally, in the special but not too uncommon situation where the borrow texture is changing frequently—that is, from day to day on a job—it is impractical to use anything other than quick, simple, low-cost methods for developing criteria for fill control specifications. (This problem, and proposed methods of solution, is presented in more detail in a following section, "Changing Borrow.")

Zero Percent Relative Density

The loosest possible practically obtainable density of a granular soil can be determined approximately by allowing an oven-dried sample of the soil to fill a container of known volume by gravity flow with little free fall. This can be done by fashioning a miniature tremie by connecting a length of rigid plastic hose to a funnel. The oven-dried soil is then placed carefully, avoiding free-fall, into a 1000-ml plastic graduated cylindrical flask. The weight and volume of the soil gives the approximate loosest dry density, from which e_L may be calculated:

$$e_L = \frac{G_s \, \gamma_w \, V}{W_s} - 1$$

where $G_s = 2.65$ (for granular soils, closely),

$\gamma_w =$ the unit weight of water, $62.4 \frac{\text{lb}}{\text{ft}^3}$ or $1 \frac{\text{g}}{\text{cc}}$,

$V =$ the measured volume, and

$W_s =$ the measured weight

One-Hundred Percent Relative Density

The densest condition can be approximated by vibrating the same sample of oven-dried soil until no further significant reduction in volume is observed. In the laboratory, this may be done most easily by securing the base of the flask to a mechanical vibrator. (An inexpensive Sepor Vibra-Pad has been used successfully.) The plot of periodic readings will establish the approximate minimum volume, and e_D may be calculated as above.

If significant amounts of clay are present, it may be necessary to pulverize dried clay lumps with a mortar and pestle before consolidation by vibration. Where large amounts of clay are present, an alternate method of establishing e_D may be adopted, as shown by Figure 4.4, which illustrates schematically the general form that compaction data would take for a TALB compacted at different energies. The quantity $\Delta\gamma$ is the difference between the maximum dry density for modified Proctor and that corresponding to 100% relative density. Research to date suggests that this increment is a relatively small number, perhaps 2–3 psf, for TALBs. (While no testing has been done, it is thought *not* to be true, in general, for other soil types, especially ECS or "pure" clays.) Thus, to establish e_D without serious error for TALBs, one could merely add 3 psf to the 100% modified Proctor value. When

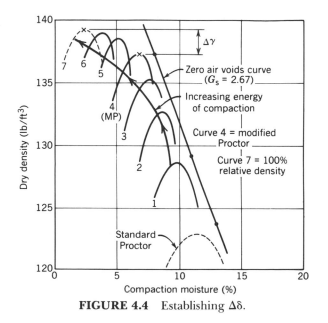

FIGURE 4.4   Establishing $\Delta\delta$.

time permits, the experimental approach of establishing $\Delta\gamma$ is recommended. As more data accumulates for a wide range of textures, the validity of assuming $\Delta\gamma$ can be determined.

Field Determination

Similar procedures may be followed in the field, as necessary. A camp stove may be used to dry the soil, and vibration may be effected manually by repeated gentle tapping of the plastic cylinder with a hammer. A small confining surcharge weight is recommended.

Percentage Field Compaction

In order to determine percentage compaction relating to the fill control target value, relative densities must be related to blow counts. Table 4.1 lists relationships suggested by Burmister.

Limitations

Figure 4.2 and Table 4.1 are applicable for foundations on sand. Extending the method to a TALB (Figure 4.3), as in the example, is therefore open to question. However, as all practicing engineers are aware, it is the essence of good engineering to exercise good judgment in the use of whatever methods are available. Accordingly, it is recommended that the method presented here is reasonably applicable to all soils that can be considered EGS, essentially granular soil, including those like many glacial tills that contain binder fractions. Where the clay is of sufficient quantity and plasticity such that the granular particles are judged to "float" in the clay matrix, the method should not be considered for use in design. The method, therefore, is definitely not suitable for clay fills. Another guiding criterion might be to consider whether intergranular contact between granular particles exist to a significant degree in the soil.

   The question of applicability is often not an easy one to answer. However, it may finally be noted that any rational method, of even limited applicability, is better than no method at all.

TABLE 4.1.   Blow Count/Relative Density Relationship of Sands (Fairly Reliable) (after Burmister)

Number of Blows per Foot, N	Relative Density	
	Designation	Percent Range
0–4	Very loose	0–10
4–10	Loose	10–40
10–30	Medium	40–70
30–50	Dense	70–95
Over 50	Very dense	Over 95

Example

A fill is to support a 10-ft wide footing and is to have an allowable bearing capacity of 4.5 tsf. What percentage modified AASHTO is required?

Look at Figure 4.2 and determine that a blow count of $N = 45$ is required. By interpolation, this corresponds to $D_R = 89\%$.

Assuming that Figure 4.3 represents the modified AASHTO compaction test on the borrow, and further that the void ratio and relative density scales have been determined as described, it is found that the required relative density corresponds to level A, which is 96.1% modified AASHTO (131.6 pcf).

Thus, the specifications for this fill could be written to require 96.1% modified AASHTO compaction. Clearly, it would then be a simple matter to determine target values for any other fills, using the specific requirements for fill support and footing size.

4.3   Changing Borrow

One of the most vexing problems in fill control occurs on jobs where the texture of the borrow being brought to the site is changing with some frequency. It is not often that borrow areas are investigated by borings, except of course for very large jobs, and so the changes are typically unexpected. When such a change occurs, it seems that a compaction test should be performed on the "new" borrow. However, if the test is to be done in accordance with ASTM procedures, a period of approximately one week is required. (See Chapter 3 for a description of the sequence of test requirements.) If the borrow area is unusually heterogeneous, the problem can become acute and virtually unmanageable because of week-long delays for compaction testing each time the borrow changes unexpectedly.

4.3.1   The Compaction Data Book

To solve this problem, I recommend the development of a published standardized document that will enable fill control personnel to *predict* modified AASHTO density within minutes for any soil exhibiting texture that qualifies as load-bearing fill: TALB or select fill. In my suggested auxiliary classification system, a GS, an EGS, and (maybe) a gemicoss would qualify, the latter depending on job circumstances such as cost, available funds, and safety. I propose that the document evolve in two stages.

Stage 1

I envision the stage 1 document as consisting of perhaps 100 grain-size-distribution curves encompassing the spectrum of acceptable textural requirements. These curves would be arranged in a standardized loose-leaf binder in logical sequence of incremental gradation changes. Included on each sheet would be the experimentally determined 100% modified AASHTO density, as determined in a carefully controlled applied research project, plus a listing of all index properties that are judged to affect significantly the compacted density of the soil. It is my judgment that these should include a uniformity coefficient $= (D_{60}/D_{10})$; Hazen's effective size, D_{10}; percentage fines; the plasticity index of fines (PI); and angularity. Also, in some cases, specific gravity differences may be a factor. A proposed format is shown in Figure 4.5.

The procedure for using the method (stage 1) would entail the use of overlays, as follows:

1. On suitable transparent plastic (Mylar) overlay sheets, plot the grain-size distribution curve for the site borrow. (Blank overlay sheets would be prepared in advance with scales identical to those of the standardized scale of the book. Time permitting, the grain-size analysis could be completed in the laboratory. Alternatively, field personnel could generate the data by hand sieving.)

2. Overlaying the plotted curve, search the book for the best fit. Note the 100% modified AASHTO density of the selected best fit.

3. To the extent judgment, experience, and confidence permits, make whatever corrections are indicated by any or all index property variations between the borrow and the selected best fit.

The last step in the stage 1 sequence is, as the saying goes, quite a mouthful. An inexperienced young person on first assignment could not be expected *at the present time* to make such judgments with any degree of confidence. I propose, therefore, that the stage 1 procedure would generally

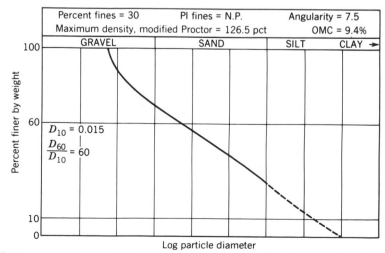

FIGURE 4.5   Proposed format of compaction data book: a specimen page.

omit any attempts at corrections, except when arrived at by, or under the guidance of, an experienced specialist. The procedure should be used, initially, only on jobs where the borrow is changing with great (and frustrating) frequency. Because of the one-week typical delay that would be required for each compaction test for each "new" soil, there really is no better alternative than to use (generally) uncorrected best fits to determine target values, especially when one considers that, with the availability of the book, the process would yield an answer in a matter of minutes.

Stage 2

I propose that stage 2 be the development of straightforward methods for determining the corrections noted in step 3 of stage 1. As an example of such a correction, it may be observed that well-graded soils have a greater potential for higher density than poorly graded soils, the former having successively smaller particles "available" for filling successively smaller void spaces. Thus, if the borrow soil has a numerically lower uniformity coefficient (poorer gradation) than the best fit, the density correction for this index variation would be negative. What the specific numerical value of the correction would be is part of the investigation of stage 2. Inasmuch as the determination of methods for assessing all corrections would require a major research effort or a series of investigations that could take many years to complete fully, I believe the sensible solution is to develop the method in the "stage approach" described. Actually, stage 2 would have a set of substages, in that when a method has been established for correcting any one of the index variations, it could be immediately incorporated as a refinement to the book.

Summary of Recommendations

What I am suggesting and recommending is that the refinements could come one at a time, over a period of many years, or all at once. I believe that the latter could occur only if a major research project was funded and undertaken at some large university with a strong geotechnical program, involving a team of graduate students directed by a bright energetic professor with strong interest, experience, and expertise in geotechnical practice and applied research.

I offer the foregoing research ideas and also appeal to some agency or group of practitioners to consider funding the projects. Stage 1, the development of the compaction data book, without correction features, is, I believe, a relatively modest and worthwhile undertaking, particularly worthwhile because it "solves" the problem of changing borrow.

Stage 2, a much more ambitious undertaking, if completed to a demonstrated high degree of reliability, would completely eliminate the need for compaction testing for all time and for all compacted fill jobs (admittedly perhaps too optimistic).

Chapter 5 contains additional commentary and recommendations regarding aspects and details of applied research and development with which I have been involved over a period of 25 years.

4.4   Problems Evolving from Traditional Practice

Many of the problems in this category have already been presented, as they could not be logically separated from other more explicit problems, so this section serves partly as a summary. Included also, however, are some suggestions for solutions that I feel would be of benefit to all concerned parties in earthwork aspects of engineered construction.

4.4.1   The Roles of Nonspecialists

Case Studies 1 and 2 illustrated the role and related problems of two very important nonspecialists: the structural engineer and the earthwork contractor.

The Structural Engineer

The basic cause of problems related to the structural engineer's functions is tied to the simple fact that the responsibility for structural drawings are logically those of the structural engineer, *including the foundation plan.* (See especially Case Study 2.) Included on the foundation plan are foundation notes. There is, therefore, a routinely created situation, frequently exacerbated by time and cost pressures, wherein the temptation exists for the structural engineer to write the foundation notes and any needed related (separate) specifications dealing with soil compaction.

My basic recommendation is this: structural engineers should not write foundation notes or other soil compaction specifications independent of advice from soil specialists. (I don't design high-rise buildings!) Alternatively, the structural engineer should undertake a reasonably intensive study of soil compaction and fill control so that, in effect, he or she becomes a specialist in this one very important area of geotechnical practice. One step toward achieving the latter objective would be the inclusion of substantial coverage of these topics in undergraduate courses in soil mechanics, and that is my secondary recommendation. This procedure would provide a partial solution for graduating engineers of the future, and pertinent parts of this book will, hopefully, serve the needs of those structural engineers already practicing.

I have always maintained that good engineering education should include the topic "when to consult a specialist," or "knowing what you know."

Construction Engineers and Contractors

The foregoing commentary and recommendations will be helpful also to construction engineers and contractors since success in bidding, planning, and supervision are inextricably linked to the ability to assess and solve earthwork problems, including interpretation of geotechnical data, boring logs, and excavation and compaction specifications.

Architects and Owners

Although considerably removed from the details of earthwork operations, the necessary interactions between the architect and the structural engineer, and between the architect and the owner, would suggest that general knowledgeability would be an important asset. When structures crack, everybody gets to see it, including the owner. Chapter 2, "Avoiding Costly Blunders," is recommended for minimum knowledgeability.

4.4.2   Fill Inspection Personnel

The problem here is that, traditionally, earthwork inspection has almost invariably been entrusted to rookies; such activity is usually one of the first assignments a young engineering graduate receives. Since undergraduate curricula provide inadequate information (in my view), the young fill inspector learns on the job, usually helped by periodic advice from the experienced project engineer, but more often left to muddle through in a trial-and-error fashion. Ironically, the better young engineer is promoted to more important activities in the office about the time he or she becomes proficient at fill inspection. Thus, there exists a vicious cycle, the result of which is cracked and failed structures.

Fortunately, there is presently evolving an excellent solution to this problem, independent of any efforts of mine: the emergence of degree programs in engineering technology (Bachelor of Science in Engineering Technology). It may take some time to graduate enough technology specialists to effect a permanent solution to this problem, but the process is under way. In the meantime, it is recommended that young engineers, engineering technologists, and those presently involved directly with fill inspection become thoroughly familiar with material contained in Chapters 6–8. Chapters 7 and 8, "Compaction Specifications" and "Fill Control Procedures," are, of course, particularly applicable.

4.4.3   Relationships Among Concerned Parties: Contractor, Inspector, Engineer

The fill inspector, above all others, is in a most delicate position. On the one hand, first and foremost is the responsibility to the owner: to ensure that the work is executed in compliance with specifications. On the other hand is the desire to get along amicably with the contractor who is executing the work. The contractor, or the contractor's earthwork foreman, is the person with whom the inspector must interact on a daily basis. In a scenario that is certain to occur, the foreman has about 30 years of experience, knows nor cares little about field density testing, and regards all 22-year-old inspectors as novices. Add to this the fact that said novice knows little or nothing about filling operations, and the stage is set for some interesting scenes!

Such situations can and do occur, and can be aggravated immensely when an unscrupulous contractor is involved. In Chapter 8, I describe just such a case history—a real horror story.

The engineer (i.e., the project engineer) has accumulated the experience and professional maturity to handle effectively the situations that arise in the field, but the problem is that he or she is not in the field on a regular, daily basis. The typical project engineer may be involved in a supervisory capacity with a number of small projects, and thus is in contact with a fill inspector on one of these projects only on a periodic or as-needed basis. This may include contacts by phone, by periodic written reports (perhaps also sent to the client), and by the engineer's occasional visits to the site.

In Chapter 8, "Fill Control Procedures," I shall offer a series of suggestions on how a fill inspector might deal with various problems that may (and do) arise in relationships with all concerned parties.

4.5   Cost and Time Pressures: A Summary

"Time is money." "I need it tomorrow." I wish I had a dollar for each time these phrases have been uttered in the frenetic world of engineered construction. The complicating factors in earthwork construction are (1) the inherent uncertainty regarding subsurface conditions (with or without borings) and (2) the weather and its particular pronounced effects on soils and site conditions.

I have alluded to specific pressures associated with compacted-fill construction. These are caused by:

1. The reluctance or procrastination in calling in a soils specialist, especially for large jobs. (This is often a cost *and* a time pressure.) It is suggested, however, that the failure to do so will often lead to greater costs in the long run, not to mention the costs measured in grief, frustration, and embarrassment. Potential legal consequences, in increasing instances, are an added factor.
2. The need to decide on the degree of compaction of a fill for a wide variety of loading conditions.
3. The particularly vexing problem of changing borrow.

The foregoing are the "easy" problems to solve, since they are largely technical in nature and thus susceptible to technical solutions. The much more difficult problems are those which are rooted in traditional practice and human interactions. Chapter 8 deals more extensively with these problems. Unfortunately I have no pat answers, as I believe there are none.

4.6   Glossary

In situ.   A Latin phrase meaning "in its natural or original position." Thus the *in situ* moisture content, unconfined compressive strength, etc., would refer to the natural moisture content, the undisturbed strength, etc.

Standard penetration test (SPT).   An important field sampling test, wherein a sampler (called a split spoon sampler) is driven into the ground in a standard fashion (140 lb hammer dropping 30 in.). During the driving, blows of the hammer are counted for successive 6-in. intervals, yielding the blow count, an indicator of soil density *(in situ)*.

Tremie.   An arrangement of linked tubes, with a wide funnel at the top, used for pouring concrete into relatively inaccessible excavations; used to avoid free-fall (and thus segregation) of the concrete.

Void ratio.   The ratio of the volume of voids to the volume of solids, expressed as a decimal. Thus a soil with a high void ratio has a low density, and vice versa.

Uniformity coefficient.   The ratio of the D_{60} size to the D_{10} size, thus an index of gradation of a soil (most meaningful for a granular soil). The larger the number, the wider the gradation. Obtained from a grain-size distribution curve. See Section 5.1.3.

5

Applied Research and Development

Over the years of my interest in compacted fill technology, stimulated by early practical exposure as a rookie geotechnical engineer, I have initiated investigations into a wide variety of factors that I felt needed more study. Some of the work has been completed to the point of publication and patents, while other studies are in various stages of early development. None of the work was sponsored, funded research. The data that has been collected, then, was largely the work of former students under my supervision at the Newark College of Engineering of New Jersey Institute of Technology.

I will not present details of this work but rather only the highlights, the purpose of which is to stimulate interest in further development to the stage of practical applications, notably with respect to the proposed compaction data book.

Factors that are suggested for further study are grouped as follows:

1. Effects of specific index property variations.
2. Waste material as fills.
3. Artificial fills.
4. Effects of mechanical laboratory compactors.
5. Density gradients.
6. Geostick correlations.
7. Percent compaction (specifications) for clay fills.

5.1 Effects of Specific Index Property Variations

In order to develop the compaction data book to its fullest (stage 2, Chapter 4), it will be necessary to determine the effects of specific index property variations to a degree that will allow for numerical corrections for each variation. (Recall that "variation" refers to each variation from the "best fit" of the compaction data book. See stage 1, Chapter 4.)

5.1.1   Percent and PI of Fines

Authorities recognize that the amount of fines in an otherwise predominantly granular soil (EGS) will affect the engineering characteristics of the soil.

Not as generally recognized, at least explicitly, is that the *plasticity* of the fines (PI) will also affect the engineering behavior of the soil. For example, some of my earliest work involved experimenting with varying proportions of granulars (+200) and fines (−200). Using commercially available powdered kaolinite, a white clay, as fines, I found at least 15% (by weight) was required to impart cohesiveness or binding qualities to a standard granular (percentages retained on the 40, 100, and 200 sieves were 67, 31, and 2, respectively.) The liquid and plastic limits of the kaolinite were 60 and 30, respectively, yielding a PI of 30. The soil mixtures thus prepared were compacted at about 12% moisture in a Harvard miniature compaction mold, extruded, and tested for unconfined compression strength. (For descriptions of equipment and testing procedures, see Lambe, 1951, p. 44.) Ten percent kaolinite merely served to "coat" the granular particles. As a result, the 90–10 mixture retained free-draining qualities, and compaction was fruitless in that the added moisture oozed from the bottom of the mold upon application of energy. At 80–20 the results were, as expected, better: less oozing and an extruded sample of sufficient cohesion to run an unconfined compression test.

Using these preliminary findings as a base, a research program was initiated wherein student groups were assigned 70–30 mixtures with the 30% fines having varying degrees of plasticity: marble dust (nonplastic), kaolinite, a modeling clay, and bentonite (drilling mud).

Cartons full of data on this and other (later) projects have been accumulated but have not been carefully evaluated, nor would its inclusion here be compatible with the purpose of this book. However, a few observations and suggestions are offered regarding effects of fines in general.

Nonplastic Fines

An ideal granular soil (one which is permanently stable) is dense, well-graded, free-draining, angular, and laterally contained. (The last requirement may be illustrated by noting that excavation adjacent to and below a footing would cause the undermining of that footing by lateral flow of the cohesionless, granular soil from beneath the footing.)

The existence of nonplastic fines (clean silts) in a granular soil will create a potential for higher density, since the void spaces the silts would occupy would otherwise be filled with air or water. However, if the percentage is high enough, these fines will clog the voids and thus render the soil marginally draining rather than free draining. The detrimental effects of changed drainage characteristics is, I judge, a bad trade-off. These effects include susceptibility to frost action, potentially damaging seepage pressures, and liquefaction (a sudden and complete loss of strength, resulting in collapse of supported structures*). From a compaction standpoint, such a soil would also be difficult to compact because of its moisture sensitivity. Finally, the fines would impart no strength to the soil by what is often called "desirable cohesive binder," because of its nonplastic, cohesionless nature.

From this, one can say that, of the four recognized soil constituents (gravel, sand, silt, clay), silt is the only one that has nothing going for it.

Plastic Fines

As noted, these fines *do* impart binding qualities to the soil mass. If the fines are present in sufficient quantities and degrees of plasticity, however, the potential problems can far outweigh, or negate, any desirable binder benefits. Principal problems are potential compressibility or expansiveness, and imperviousness. The latter characteristic entails additional problems of seepage pressures (including uplift), and excess porewater pressures (reducing shear strength). Clays are also compaction-energy sensitive, and compacted embankments of high clay content are highly anisotropic, exhibiting markedly different strength and seepage characteristics in different directions.

Since compaction is inherently a lubrication process (Figure 3.1), the interaction of water with a fill containing plastic fines (or one which is 100% clay) may be thought of as "shared water." As described previously, clay particles of different mineralogies may have markedly different affinities for water. The water that attaches to the clay particle surfaces (actually called "bound water" by many authorities) has many of its properties changed by complex electrochemical activity, a fundamental cause of plasticity. One of these changes renders the bound water more viscous than normal water, and thus less capable of providing the lubrication necessary for compaction. Thus, water added to a dry clay soil can be thought of as being divided into bound water and lubricating water, the proportions depending upon how much the particular clay mineral requires for bound water satiation. For highly plastic clays such as bentonite (liquid limit about 400%), one would have to satisfy this bound water demand before any lubrication can take place for effective compaction. This reasoning makes it easy to see why optimum moisture contents for fat clays are so high compared to lean clays such as kaolinite (Figure 3.4).

---

*Contractors and equipment operators should be aware of the very real danger of equipment (and operators!) being swallowed up with very little warning if an overly vigorous attempt is made to compact thick lifts of loose, saturated sands with heavy vibratory compactors, particularly when dealing with hydraulic fills (Sowers, 1970, p. 233).

Compaction Data Book Corrections

For purposes of assessing the effects of percent and PI variations, it would seem reasonable to conclude the following:

1. The density corrections for a PI variation would be inversely related. That is, if the PI of the site borrow fines is higher than the best fit, its effect would be to lower the compacted density of the borrow, all other factors being equal.

2. The density correction for a percentage fines variation would also be inversely related; the more fines, the lower potential density.

3. Optimum moisture content corrections would be directly related to PI variations. Thus if the PI of the site borrow fines is higher than the best fit, the optimum moisture content of the site borrow would be higher.

5.1.2   Hazen's Effective Size: D_{10}

The D_{10} size was postulated by Allen Hazen to be that size that most significantly affects the seepage qualities of a soil.

Hazen's work dealt principally with sand filters in water treatment plants (i.e., sanitary engineering). He probably reasoned, quite logically, that the sizes of the smaller void spaces would govern drainage qualities, and that these void spaces would be related to comparable particle sizes. The 10% size (meaning that 10% of the soil is finer than that size—see Figure 4.5) was chosen for study, and an empirical equation was developed relating permeability to the D_{10} size:

$$k = cD_{10}^{2}$$

where k is the permeability coefficient (in millimeters per second) c is a constant between 10 and 15, and D_{10} is Hazen's effective size in millimeters (Sowers, 1979, p. 96). The equation is principally applicable to clean sands and is usually considered to give only an indication of the order of magnitude of the permeability.

A preliminary investigation has been made to determine the relationship between D_{10} and the compaction characteristics of soils bordering between freely draining soils and marginally draining soils. As illustrated in Figure 3.4, select fills (freely draining) do not exhibit a discrete optimum moisture content, and, therefore, it makes no sense to attempt a standard compaction test on such a soil. The problem is, of course, to have a way of identifying such a soil quickly by some easily determined index property. It was the purpose of my investigation to determine the approximate D_{10} size that identifies a select fill as such, and that is one of my suggestions for future applied research and development. (As noted in Chapter 1, Section 1.4, select fills are sometimes defined as soils containing not more than 12%

passing the No. 200 sieve. This would serve as a good starting point for the investigation I propose.)

5.1.3　Uniformity Coefficient: D_{60}/D_{10}

A well-graded soil is one which contains successively smaller particles filling successively smaller void spaces, thus creating a potential for greater density. The uniformity coefficient D_{60}/D_{10} is a number or index that reflects the gradation of the soil. (See Figure 4.5.) Clearly, the larger the uniformity coefficient, the wider the gradation. Thus, if the site borrow has a greater uniformity coefficient than the best fit, the compacted density of the site borrow should be greater.

No research has been done (to my knowledge) to determine a specific relationship between uniformity coefficient and compacted density.

The uniformity coefficient is a standard index property in the Unified Soil Classification System. Thus, the research I propose would have immediate and easy applicability in practice.

5.1.4　Angularity

Geologic processes influence greatly the angularity of predominantly granular soils, particularly the abrasion effects during transport by water or wind (alluvial and aeolian soils, respectively). In general, highly angular soils, if dense, are extremely strong and stable because of the interlocking of particles. Angular particles also exhibit rough surface texture, like sandpaper, so that particles interlock also in a surface-to-surface fashion. Thus, interlocking may be thought of as occurring on a macroscopic and microscopic level. Means and Parcher (1963) contend that the angularity of granular soils has a much greater effect on the soil's behavior than the paucity of investigations suggests, and I concur.

One of the first steps in such an investigation would be the development of a practical method of assessing a soil's angularity quantitatively, analogous to the hardness number of minerals. Thus a soil might have an angularity rating between, say, 1 and 10 for spherical and pyramidal shapes, respectively. Once such a rating method is established, it would be possible to study angularity effects on compaction and other engineering properties such as shear strength and compressibility.

I have supervised some work along these lines. As a first step, the literature was searched to determine if any attempts had been made by other investigators to quantify and study angularity. As I suspected, little if any such work had been done. Indeed, all textbooks seem to classify granular soils as angular, subangular, subrounded, and rounded and let it go at that.

Angularity effects appears to be a neglected area of study.

5.1.5   Specific Gravity

Means and Parcher (1963) suggest that the specific gravity of granular soils, which would be composed primarily of quartzitic minerals, can be taken as 2.65 for most practical applications—that of clay minerals somewhat higher, about 2.70. It is evident, however, that some instances will arise where "peculiar" mineralogy will dictate that the assumption of a specific gravity of 2.65, 2.70, or an intermediate value for a mixed soil, will introduce unacceptable error. Fortunately, with experience, one can easily identify situations where specific gravity variations may be significant. There are two ways in which this can be done. First, the soil feels heavy (or light), and it is surprising how this simple tactile technique allows one to detect significant specific gravity variations. For example, the specific gravity of cement is 3.1, and even an inexperienced person can easily detect the "heaviness" of a handful of dry cement compared to, say, a handful of powdered bentonite (drilling mud). Try it! Second, one can develop an intuition for the "correct" maximum conpacted density at a given level, as suggested by Figure 3.4. When the answer is suspect, it may be the result of significant deviation of the specific gravity from the usual 2.65–2.70 range.

When such deviation is suspected, the specific gravity should be determined. I would, however, offer a note of caution: Be sure to have the work done by a company whose laboratory work you have learned to accept as accurate. It is my opinion that the classical, direct method of determining the specific gravity of soil minerals is unusually susceptible to experimental error, so much so that assuming an answer is often more accurate than trusting an experimental result, particularly when experienced persons apply their judgment to the specific case.

I have seen data that suggest that only the very best laboratories can perform this "simple" test satisfactorily. Sometime around 1960, the American Council of Independent Laboratories conducted a Standard Soil Samples Program, which involved sending soil samples to 99 participating laboratories for certain routine soils tests, including specific gravity determinations and modified AASHTO compaction.

Three "umpire" laboratories, presumably chosen for their reputations and prestige, were chosen to do the same tests. Their results established the control (i.e., correct) values.

Following are the maximum, control, and minimum values reported for a CH soil (clay of high plasticity, Unified System) (Hirschfeld, 1965):

	Maximum	Control	Minimum
Specific gravity	2.79	2.70	2.21
Modificd AASHTO	123.9	114.0	105.4

And, mind you, since these were participating laboratories, they undoubtedly put forth their very best efforts.

To some degree, this cautionary note pertains to any laboratory testing you may need. Where very large jobs are involved, "second opinions" may be warranted.

5.2   Waste Materials as Fills

As urban and industrial development occurs in a given region, good fills within reasonable haul distance become more scarce and expensive. At the same time, the very processes of urbanization and industrialization generate larger and larger quantities of what was once considered waste but is now being viewed as recycleable material. Among these products are ash, glass, rubber (tires), aluminum and other metal containers, and a broad category one might call construction rubble.

In addition to the more common waste materials, there exist throughout the world a variety of materials of which there is an annoying abundance, materials such as sulfur, bamboo, and, of course, garbage. An interesting publication that deals with the use of such materials in construction is *New Horizons in Construction Materials* (Fang, 1976). This publication has served as a basis for further research investigations in my work, and I recommend it for that purpose for other investigators, including students searching for interesting, challenging, and unusual thesis topics. Some will be particularly suitable for persons from the regions where the materials are abundant, for in many cases standard materials are unavailable or too expensive. In most cases, the uses of the materials may be extended to the construction of fills or their reinforcement. Earth fills reinforced by bamboo is one suggested possibility. Obviously, a certain amount of research will be necessary to bring such technology to construction applicability.

Chae and Gurdziel (1976) have studied New Jersey fly ash as a fill, and report design strengths of about 5 tsf. One of my investigations involved mixing their fly ash with natural soils of better texture to determine the engineering and compaction characteristics of the mixture. Obviously, various mixture proportions will produce fills of differing properties. Since the fly ash has normally been simply discarded (at some hauling cost), its use as a fill or fill component is a potentially attractive economic alternative.

With my encouragement and supervision, students have investigated the use of glass and rubber as fill components.

Sulfur is being investigated as a replacement for asphalt in bituminous pavements, a potentially enormous cost reduction (McBee et al., 1976).

As is now well known, the use of garbage in landfills is common. Indeed, with the advent of the environmental movement, and the horrors of Love Canal, what goes into "sanitary" fills is now of major concern and study. Believe it or not, the Fritz Laboratory of Lehigh University once did a compression test on one cubic yard of compacted raw garbage (which arrived in July from California, undoubtedly to the dismay of receiving department personnel). A photograph of this "first" appears on p. 133 of *New Horizons in Construction Materials.*

5.3　Artificial Fills

In 1970, I was struck by the idea of using foam plastic as a complete replacement for soil as a structural fill. A major benefit of this would be a dramatic weight credit. This simple concept resulted in the acquisition of two patents on the methods (Monahan, 1971, 1973).

5.3.1　Weight-Credit Foundation Construction Using Foam Plastic as Fill

If a hole is dug and the material that is removed is immediately used to fill the hole, the state of stress below is unchanged. It follows that if a lightweight backfill is used, there will accrue a weight credit for a proposed structure. In an extreme case, if the weight of the backfill plus the structure is equal to the weight of the soil removed, then no settlement of the structure can occur. This principle is used extensively in Mexico City to produce so-called floating foundations on the highly compressible volcanic clays underlying the region.

In less dramatic cases, lightweight backfills (e.g., cinder fills) may be used to effect a lesser weight credit, thus reducing the net stress increase on compressible materials below.

Apparently because of the scarcity of natural lightweight aggregrates and the consequent infrequency of fortuitous circumstances that might dictate its use, the application of the technique has not been extensive.

I propose the use of foam plastic as backfill in certain special cases. Technology exists or can easily be developed for on-site production of the foam plastic, or alternatively, the material can be precast for installation at the site.

Illustration of Weight Credit
Consider a site that is overlain by compressible material of soft consistency, for example, a swamp muck.

Unit weight of muck	100 pcf
Depth of excavation	15 ft
Stress release	1500 psf
Foam density	3 pcf (typical)
Stress addition	45 psf
Stress credit	1455 psf

It is common that in areas where soils of such marginal supporting capacity exist, the typical methods of soil stabilization (e.g., preconsolidation with or without sand drains or the use of compacted sand backfill) are suitable for developing an improved condition to allow building pressures of about 500 psf, corresponding to one-story light industrial buildings or office buildings. It is seen that the stress credit in this example is about three times this value. And this trebled allowable building pressure could theoretically be applied without causing any settlement.

The compressive strength of a foam which has been used for such an application (Styrofoam® HI) is 35 psi. Thus, the building pressure of 1425 psf (±10 psi) would stress the plastic to a level of less than one-third of its compressive strength.

A perhaps oversimplified conclusion is that a three- or four-story building would now be possible where only one-story construction would be technically feasible using standard techniques.

Foam Materials

The only foam plastic that is known to have been used in weight-credit engineered construction is a (solid) precast extruded polystyrene foam designated Styrofoam HI, one of approximately 12 Styrofoam materials made by the Dow Chemical Company. The Styrofoam HI was used on the construction of the Pickford Bridge in Michigan (Coleman, 1974).

To illustrate the broad potential for the general use of foam plastics (precast and blown-in-place) in weight-credit construction, consider the following quotation from *Modern Plastics Encyclopedia* (1968), "Urethane Foams," p. 346:

foams may be produced which have densities ranging from less than one pcf to about 70 pcf, with an almost limitless range of chemical and mechanical properties.

In light of the above, it would seem feasible that for very large jobs, or very special circumstances, one could justify the expense of producing a foam plastic of special formulation to suit the particular use. In addition, one could use combinations of existing products (i.e., Styrofoam products of different properties) to effect a design. As an example, one could use a high-quality Styrofoam immediately beneath a footing where stress levels are relatively high and a lower-quality (and presumably cheaper) foam plastic below. This would be analgous to the manner in which pavement systems are designed.

Table 5.1 lists technical data on the specific Styrofoam product that was used on the Pickford Bridge, where a wedge-shaped block of "bundles" of Styrofoam HI-35 boards were installed. Figure 5.1 illustrates the construction schematically. The soft clays were so weak that the abutments (prior to weight-credit construction) were tied to trees for added stability! With the use of Styrofoam fill, the maximum pressure (at the rear of the abutment

TABLE 5.1. Properties of Styrofoam HI-35 (from Dow Chemical brochure)

	Test Data	Test Method
Compressive strength at 5% deflection	35 psi (2.5 tsf)	ASTM D1621-59 T
Water absorption	0.25% (by volume)	ASTM C-272-53
Density	2.5 lb/ft^3	

walls) was reduced from about 1200 psf to about 612 psf (5 ft of soil fill at 120 + 5 ft of Styrofoam at 2.5). The soft clays were able to sustain this reduced pressure without detrimental settlements.

The 5 ft of soil fill was for two purposes: to "pin down" the foam plastic to avoid possible flotation and to minimize the amount of relatively expensive foam plastic ($40 per yard).

I was informed by the Dow engineer involved with the project that the subgrade was so soft that one would sink to his knees when attempting to traverse the area with normal footwear. The bundles of Styrofoam were placed directly on the untreated subgrade, and two men placed one side in one morning.

A sheet of polystyrene was placed on top of the Styrofoam to protect against the possibility of deterioration of the plastic by spillage of gasoline in the unlikely event of a tank-truck crash at the approaches to the bridge.

The water absorption characteristic of the material used for a weight-credit application is important, inasmuch as water pickup will result in weight increase. To quote a prominent engineer whom I consulted: "I hate to bury voids" (Johnston, conversation, 1976). As noted, the water absorbtion of Styrofoam HI is virtually negligible (0.25% by volume).

Table 5.2 contains strength–weight ratios that illustrates by contrast this unusual property of rigid foam plastics.

Permanency, Durability

The most convincing evidence of permanency and durability are those instances of field usage.

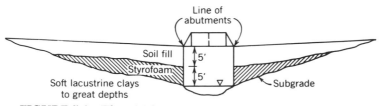

FIGURE 5.1 The Pickford Bridge crossing, Pickford, Michigan.

TABLE 5.2.   **Strength–Weight Ratios**

	Compressive Strength tsf Unit Weight (pcf)
Gravel (exceptionally compact)[a]	$\dfrac{10}{130} = 0.08$
Bedrock (massive, sound)[a]	$\dfrac{100}{170} = 0.59$
Styrofoam HI-35[b]	$\dfrac{2.5}{2.5} = 1.00$
Reinforced concrete	$\dfrac{217}{150} = 1.45$
Styrofoam HD-300[b]	$\dfrac{8.64}{3.3} = 2.70$

[a]Various building codes; typical.
[b]Dow Chemical Company.

The Pickford Bridge.   This is the only known instance of large-volume use of foam plastic in a weight-credit construction application. As reported to me orally, there was an immediate settlement of about 2 in.—apparently an adjustment of the stacked bundles of Styrofoam—but no subsequent movement of significance has been observed since.

Highway Insulation.   Since 1962, Styrofoam HI has been used in at least 39 installations as insulation for highway pavement systems. In most cases, the amount of plastic used has been about 1 to 3 in. (typically placed directly on the subgrade). Samples of the foam taken from various highways after several years of service have shown very little moisture pickup (Williams, 1968).

Structural Performance.   The structural performance of a highway insulated with Styrofoam HI was investigated by personnel of the Maine State Highway Commission (1966). Deflections were of the order of 0.02 in. in the insulated sections and were well below those of the control section during the spring thaw, the latter reaching a maximum of 0.03 in.

Utility Installations.   Thirteen utility installations were insulated with Styrofoam between 1964 and 1966. These included sewer lines and water lines. No case of a pipeline application involving the purposeful use of foam plastic *for weight credit* is known.

Pole Installations.   A foam plastic compound called Poleset® is being used to install poles in the ground. The injected compound is used in lieu of the usual technique of compacted soil backfill. Users, notably power companies, report in their literature that the material is stronger than the *in situ* soil, based upon pulling tests in the field. In such a usage, weight credit is not a factor. Apparently, the scarcity of select borrow, the neatness of the operation, and reduced labor costs combine to make the technique workable and economically feasible.

Laboratory Tests.   Accelerated laboratory tests such as freeze–thaw cycling and soak tests have shown very little increase in water pickup (Williams, 1968).

Cost and Cost Analyses

The in-place cost of Styrofoam HI at the Pickford Bridge was $40/yd. This was high as compared to the cost of compacted fill in-place (perhaps $8/yd). However, a comparison of unit costs is not usually valid in appraising the feasibility of the weight-credit method. Instead, I suggest the following approaches.

Real Estate Context.   Perhaps the best context in which to gauge the worth and economic feasibility of a particular application is to compare the cost of stabilization (or *improved* stabilization) with that of the real estate value. For example, experts have estimated the value of lower Manhattan real estate at $400–500 per square foot (about $16 million per acre). The cost of placing Battery Park fill is estimated at about $14–15/ft^2, with a $40 upper limit when utilities and other costs are considered. Thus, it is clear that "making" this new ground (93 acres) was a tremendously sound investment.

In areas such as the New Jersey meadows, where the land is also inherently valuable (in a real estate context) because of its geography, but is undeveloped because of geologic (foundation) shortcomings, a similar type of cost comparison would be valid. Thus, if a foundation construction technique can yield a three- or four-story capability as opposed to the present one-story capability, the client gets three or four times as much floor space.

Total Cost Context.   In many cases, the best approach is to look at the bottom line or total cost of a job and compare that cost to other "traditional" construction methods. A Staten Island highway project demonstrates this approach. (See *Highways* following.)

Possibility Context.   In some instances, it may develop that no construction is possible without exceptional weight credits afforded by the use of foam plastic fill.

Hypothetical Case Histories

In order to illustrate further the application of the weight-credit technique, a number of hypothetical case histories follow. Some are general and merely conceptual; but some suggested applications are for the solution of specific, real problems in foundation construction.

Route Construction in General.   If a route of any type (highway, railroad, pipeline) is to be constructed between two points A and B and a large area of marginal soil exists between these points, the usual solution would be to go around the area (see Figure 5.2). With the dramatic weight credit that can be attained by the use of foam plastic (as an artificial subbase for a highway, for example) it is conceivable that the marginal soil might be crossed. The length of the route would thus be shortened and the resultant savings might make the straight route economically superior. Possibly, the loads imposed by rail equipment would be so high as to preclude its application to railroads, but highway and pipeline loadings seem within the scope of application. When it is considered that interstate systems cost in excess of $1 million/mile (depending, of course, on cuts, fills, locale, etc.), the cost of the plastic foam and its installation could be justified.

Highways.   Healy (1975) cites a highway project that illustrates both the method and the manner of appraising in a total cost context:

A recent example of a case where foam plastic backfill would have provided a better and less expensive solution to a soil problem was the construction of a major arterial highway in Staten Island, New York. The author was directly involved with this project which bisected a tidal marsh. The method specified in the contract for subgrade stabilization was to excavate unsuitable material and backfill with compacted 1½ inch broken stone. The purpose of using 1½ inch broken stone instead of sand fill was to have larger voids, thereby generating a larger weight credit. The contract unit price for stabilization was $33.00 per cubic yard—$13.00 for excavation and $20.00 for in-place 1½ inch broken stone fill.

The subgrade was so poor that it was necessary to excavate as much as 10 feet in certain areas. The final excavation quantity for stabilization was 83,350 cubic yards which generated a total stabilization cost of $2,750,550.00. However, if foam plastic was utilized as backfill for the purpose of gaining weight credit, as was suggested to the New York City Highway Department during the early stages of the project,

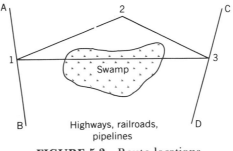

FIGURE 5.2   Route locations.

a substantial savings would have been realized. Only 10% of the final excavation quantity (8,335 cubic yards) would have been required to achieve an equivalent weight credit. The total cost for stabilization if polyurethane was utilized would have been approximately $441,755.00 (8,335 cubic yards at $13.00 per cubic yard for excavation plus $40.00 per cubic yard for foam plastic backfill).

Therefore, it would seem that a savings of about 2.31 million dollars could have been achieved if foam plastic backfill was used. In addition, foam plastic probably would have produced a better overall project.

Pipelines.   The use of foam plastic as an injected backfill under and around pipelines is thought to be practical (see Figure 5.3). Envisioned is a type of wheel-mounted machine that would move in a straddling fashion along the trench, producing and injecting plastic foam into the trench. This application might be particularly appropriate in large, congested cities for two reasons. First, the existing pavement would provide the needed support for the machinery, and second, the trucking of soil fill to the site (through congested city streets) could be eliminated.

In the case of fluid in a pipeline that is designed for gravity flow, it is apparent that the size of the pipe wall can be reduced to that of essentially a form since the only reason for a pipe of any substantial strength in such a case is to withstand the pressure of the backfill and any live loads. Since the pressure exerted by a foam plastic fill would be negligibly small, the required strength of the pipe would be governed principally by live loads. In fact, it is speculated that a slip-form of some type could eliminate the need for a pipe; the slip-form could be moved forward after the plastic foam has created a conduit through which the fluid would flow.

Thus, it is conceivable that, for the case of urban pipeline construction, the costs of soil fill, its transportation, and the costs of the pipe (gravity flow) could be reduced or even eliminated (i.e., weighed against the costs of the foam plastic weight-credit approach.)

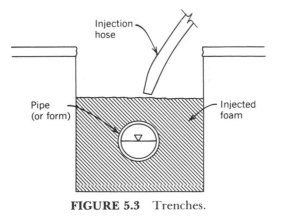

FIGURE 5.3   Trenches.

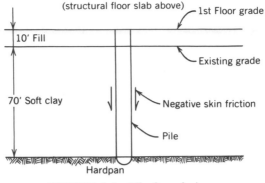

FIGURE 5.4   Pile foundations.

Because of the heat insulating properties of foam plastics, its use to encase the oil pipelines through permafrost is suggested. Where needed, weight credit and insulation could be achieved at the same time.

A Pile Load-Test Program.   I worked on a pile load-test program in which steel H-piles were driven to bedrock to depths of about 70 ft, with specifications for equipment and driving to produce a design pile of 175 tons (see Figure 5.4).

Above the bedrock were thick deposits of soft clay. Finished first-floor grade was at a level several feet above existing grade; and the design called for the space between existing grade and the slab to be a fill material. The high floor loads dictated a structural floor slab, so the only reason for the soil was to fill the space under the building. The weight of this fill would, however, cause the consolidation of the thick, compressible clay below, and it was estimated that the drag on each pile caused by the settling clay would be about 50 tons (negative skin friction).

The use of a foam plastic fill could have eliminated the substantial effect of drag on the piles, and smaller piles could have been used to support the building-imposed pile loading of 125 tons.

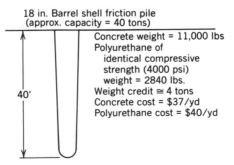

FIGURE 5.5   The Healy pile.

The Healy Pile (Figure 5.5). Another excerpt from Healy (1975) follows:

The weight credit obtained if polyurethane foam plastic was substituted for concrete in piles would be substantial. The pile bearing capacity could then be increased by the resulting weight credit.

For example, an average 40 ton capacity, 18 inch diameter, barrel shell friction pile which would probably be driven to a depth of about 40 feet, would have a volume of approximately 71 cubic feet. If concrete was utilized to fill the pile, it would weigh approximately 11,000 pounds. On the other hand, polyurethane plastic, which weighs 40 pounds per cubic foot for an equivalent compressive strength (4000 psi), would yield a pile weight of 2,840 pounds; weight credit, about 4 tons. As a result the capacity of the pile could now be considered as 44 tons, an increase of 10%.

In general, the weight credit generated by substituting foam plastic in any deep foundation system which utilizes concrete, would produce significant increases in allowable bearing capacity. In addition, it should be noted that the cost of materials are approximately equal — 4,000 pounds per square inch concrete costs about $37.00 per cubic yard and 4,000 psi polyurethane foam costs about $40.00 per cubic yard.

A Grade-Separation Case Study (Figure 5.6). I became aware of plans to effect a grade separation at an important intersection in a major eastern city. In discussions with the chief soils engineer it was learned that the principal complication of the design was that a subway ran directly beneath the intersection at a relatively shallow depth. And the subway at that point is presently loaded "about to its limit," so the approach fills necessary to make the grade separation create a problem of overstressing the structural elements of the subway system.

The solution was to excavate some of the material above the subway and backfill to the planned elevated grade with lightweight fill. I inspected the fill and it appeared to be some form of ash material containing considerable black silty material, a rather poor fill by normal standards (poor drainage, difficult compaction, probably frost susceptible), but its unit weight was about 65–70 pcf. Thus, a weight credit would be attained and apparently the material was acceptable for the solution of this special problem.

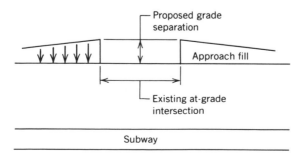

FIGURE 5.6 A grade separation.

The use of a foam plastic approach fill (at least in part) could have provided a much greater weight credit and would have the advantages of ease of construction and elimination of drainage and frost problems.

The Catskill Aqueduct Case Study (Figure 5.7). Within a 48-mile section of the Catskill Aqueduct System there are included approximately 15 miles of pressure tunnels. This construction was used to cross filled valleys where no suitable material was available to support other kinds of construction. In general, the depth of these pressure tunnels was governed by the depth of preglacial gorges that contained glacial drift and other recent unconsolidated alluvium. The deepest pressure tunnel was for the Hudson River crossing; it was founded at a depth of 1111 ft below the level of the river. The length of the tunnel at this depth is over 3000 ft.

No cost figures are available for the construction of these pressure tunnels, but the figures must be impressive for such an effort. Had the technology been available, it is probable that great savings might have been possible by using plastic foam backfill to "float" across the preglacial gorges with cut-and-cover construction.

Airport Construction. Foam plastic subbases might be used to establish a weight credit in airport pavement design. Because many airports must be located in areas of marginal soil support—bay muds, for example—the technique may have wide applicability. (It may be noted that most stream crossings, as at the Pickford Bridge, would also involve construction of approach embankments of soft or loose sediments because of recent alluvium).

Retaining Structures (Figure 5.8). The uses of foam plastics behind retaining structures is a logical extension of the weight credit concept. In this case, the lateral pressure on the retaining structure caused by the pressure of the backfill itself will be reduced to a negligible value with the use of the foam plastic in place of the usual soil backfill. Thus, where no major

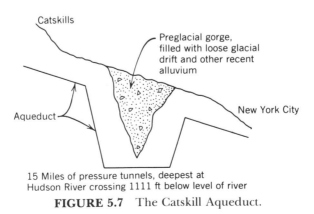

15 Miles of pressure tunnels, deepest at
Hudson River crossing 1111 ft below level of river

FIGURE 5.7 The Catskill Aqueduct.

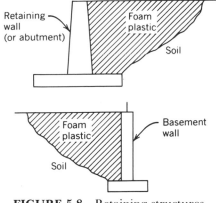

FIGURE 5.8 Retaining structures.

live loads are involved, the size and strengthening of the wall required to retain the backfill will be reduced substantially. Drainage design can also be greatly simplified.

A possible application would be the use of wedge-shaped sections of pre-cast foam plastic for various size excavations adjacent to basement walls. Or a system might be developed to fabricate or inject the foam on site. For large construction projects, the latter technique would be preferable.

Slope Stability. Figure 5.9 (Wager and Holtz, 1976) illustrates a solution to a difficult slope stability problem through a unique use of reinforced earth technology. The forces P and $P \tan \phi$ providing counterclockwise resisting moments are developed by the inclusion of short sheet piles that are connected by pretensioned steel, the steel being stressed to provide a factor of safety of at least 1.5. This procedure, while unique and effective, has something in common with most approaches to increasing slope stability: providing greater resistance to failure.

It is my opinion that many slope stability problems can be solved at less cost by considering the "neglected part" of the factor of safety, the driving

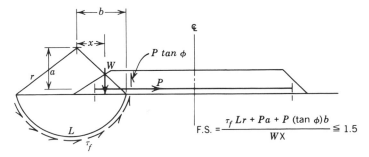

$$\text{F.S.} = \frac{\tau_f \, Lr + Pa + P \, (\tan \phi)b}{WX} \leq 1.5$$

FIGURE 5.9 Slope stability. (From Wager and Holtz, 1976.)

forces—in this (and most) cases, the weight W. The use of some foam plastic in the embankment could reduce or perhaps eliminate the need for re-inforcement. In cases where elimination of reinforcement can be achieved, the savings in construction labor costs could be major.

Technical Problems—Some Suggested Solutions

The presently recognized technical questions relating to the use of foam plastics in weight-credit applications to foundation construction are related to permanency and durability under the influence of load and exposure to natural and man-made environments.

Special problems during service relate to differential icing, chemical resistance, long-term compression, and flotation.

Differential Icing. On highway insulation installations, the purpose of the foam plastic was to prevent or minimize frost action in susceptible subgrade soils (i.e., silts). Experience has shown that the insulation can protect the subgrade, but in some instances can cause the pavement to become icy along treated sections of the roadway, thus creating a hazardous driving condition.

Solutions could include avoidance or special designs. Also, where differential icing is considered to be a possibility, special caution signs may be used, as are customary at bridge approaches.

Avoidance. It is my opinion that the extensive use of foam plastic under long sections of high-speed highways in cold regions should be avoided (or very carefully considered) because of the potential hazard of differential icing.

Dow Chemical studies suggest that differential icing will not occur in regions where the cumulative degree-days (°F) do not exceed 100. This would include most of the southern-tier states.

While each case of potential highway usage should be considered on the basis of its specific technical problems, economics, and risks (of differential icing), it is thought that the procedure would be reasonable in cold regions for short sections of low-speed roads and (especially) for grade separations (as at the Pickford Bridge) where the loadings of approach embankments can be severe with normal fills. In such cases, a sign can be posted such as CAUTION—PAVEMENT MAY BE ICY NEAR BRIDGE. Such a sign would be but a slight modification of present practice warning of possible icy pavement *at* bridges.

A case of reasonable usage for a low-speed road involved a short access road to an expanded golf course in Connecticut. Apparently because the project was an expansion of an existing facility, there was little or no choice for the location of an access road, and the area was underlain by very soft soils to considerable depths. I was advised that this road section was re-peatedly surcharged (i.e., repaved) to the point where the trees on both sides became visibly tilted. The cost of such periodic maintenance was substantial and probably far exceeded what the cost of floating the road (with the use of foam plastic) would have been.

Special Designs.   Where circumstances dictate, special designs could include "heat bridges" to allow enough heat to leave the subgrade to prevent icing, yet provide the insulation necessary to protect the subgrade from frost action: an optimization approach. Aluminum pipes or sheets are suggested as heat bridges.

Computer studies by Dow Chemical have been developed that enable the computation of sufficient soil cover *over* the foam plastic to provide the necessary heat sink to preclude icing.

Chemical Resistance.   Foam plastics (e.g., Styrofoam) generally have poor resistance to some materials, most notably gasoline. As noted for the Pickford Bridge construction, polyethylene sheets were used to cover and protect the Styrofoam bundles against possible gasoline spills.

The possibilities of exposure to other materials should be considered for each case and for each type of foam plastic used. Suppliers can supply detailed information and recommendations.

Long-Term Compression.   A question that arises regularly in discussions with engineers who are considering the method is that of long-term compression in service—creep. Most studies have been done on the relatively thin sections of Styrofoam HI that have been used for insulation, as presented earlier, and the permanence and stability would seem favorable.

Perhaps the best (and only) evidence of stability that generates confidence for weight-credit applications (where greater thicknesses would be used) is the Pickford Bridge performance. As far as I am aware, no problems have developed.

Flotation.   At Pickford Bridge, 5 ft of soil fill, placed above the plastic, was sufficient pressure to "pin down" the foam material in the event of flooding and the accompanying rising water table.

Soil pins, analogous to anchor bolts in tunnel construction, might also be employed where larger weight credits are needed.

In cases of support of buildings (as opposed to embankments, pipelines, etc.), flotation problems would not generally be acute; the design approach would be to create a positive pressure in the subsoils consistent with (their) allowable pressure tolerances. In general, zero stress increases would be avoided.

Sunlight.   Direct exposure of foam plastic to sunlight for extended periods of time shoud be avoided.

Future Usage

It is my feeling that injected (cast-in-place) foam plastic would hold great promise for future usage because of several factors. First, the materials that produce the foam can be brought to the site in appropriate containers of much smaller volume than that of the (ultimate) injected foam plastic. Second, the materials can be injected into cavities of any shape. Third, there is a certain "neatness" to the operations envisaged that suggests major savings in labor costs. Finally, such backfills will eliminate many of the problems of soil filling and compaction: texture (grain-size) specifications and compaction and fill control, especially in confined spaces such as trenches and wedge-shaped fills behind retaining structures and basement walls. (It is my opinion, based upon many experiences and much reading, that such backfilling procedures are very often done incorrectly and lead to many problems of a technical and "litigious" nature.)

On the potentially negative side, it is thought that foam quality and quality control would be less, as compared to precast foams such as Styrofoam. Compressive strength and water absorption would have to be considered carefully and estimated conservatively. With respect to water absorption, usage might be often restricted to areas above the water table, to retain weight credits throughout the life of the structure.

Precast foams will have most usage where quality control is vital and where consequences of distress or failure are major. As with any new method, confidence in the method is necessary for extensive future usage. Usage with success will breed routine usage.

5.4   Effects of Mechanical Laboratory Compactors

Anyone who has ever done a modified Proctor compaction test alone and by manual means will appreciate the assertion that it is hard, boring work. Assuming five points would be needed to define the compaction curve (i.e., optimum moisture content), the 10-lb hammer would have to be lifted 18 in. 625 times. Because of the hard physical work and the tedium, there are opportunities for error from several sources: not lifting the hammer the full 18 in. (especially as fatigue sets in); not executing a vertical free-fall, and thus reducing the energy of impact through side friction between the handle and sleeve of the hammer; slinging the hammer (especially if the radio is on to lively rock music!); losing count (blows or whole layers); and differences in coverage patterns among individuals.

Apparently recognizing a need, equipment manufacturers have developed a mechanical compactor that can be set to deliver automatically a given number of blows from a given height, with provision made for interrupting the action for the addition of successive layers. While it is apparent that human errors are rectified by using the mechanical compactor, ASTM procedures allow for either manual or mechanical means of compaction. It is my belief that this optional feature of the test procedure is unfortunate, in that major differences in results can occur because of variations among individuals opting for the manual technique.

I am not aware of any studies to determine differences between hand compaction and machine compaction. Such an investigation would require an extensive amount of work, for we are here dealing with human variances. Many tests would be needed to establish statistically valid conclusions. Also, we have a Catch-22 situation: If ASTM was to eliminate the option by requiring machine compaction (a step that I recommend), a study would be needed to correlate future test results (by machine) with past results (mostly manual). The alternative would be to simply regard all past results as worthless.

Finally, another unfortunate feature of present ASTM procedures is that the tester does not even have to stipulate (on the data sheet) which option was used for compaction, thus rendering the work worthless because of uncertainty on the part of the potential user.

5.5   Density Gradients

When a soil is compacted in the laboratory (or in the field), the lower layers (lifts) will be compacted to a higher density because the zone of influence of the hammer (roller) exceeds the layer (lift) thickness. This being the case, it would seem reasonable to pass a field fill even if the field density test value is less than the target value, as long as the test value is within an established range, as determined by a carefully conducted research investigation of density gradients.

5.6   Geostick Correlations

The Acker Drill Company makes a device called a geostick, which is a combination penetrometer, geologist's pick, and field sampler (for soft soils). The penetrometer feature allows one merely to push the conical tip vertically into a soil either under the weight of the stick for soft soils, or with the weight of the operator for stiff or dense soils. Based upon the penetration values, a presumptive bearing capacity is read (or usually interpolated) from a table printed on the barrel of the device.

I have begun very preliminary studies to attempt to correlate geostick readings with compacted densities, and to investigate density gradients.

Since the zone of influence, or pressure bulb, for the conical tip is at most a couple of inches, such readings should never be considered a complete substitute for direct means of field-density testing. Also, the proximity of the rigid wall of the laboratory mold may significantly complicate correlations between laboratory and field readings. However, I feel that the technique could serve as a valuable additional aid in fill control work, as long as reasonably valid correlations are developed and good judgment is used in recognizing its limitations and uses. For example, one would be foolish to arrive at a site and pass a 10-ft thick embankment based upon a geostick reading at the surface. If one, however, has *observed* the compaction process, lift by lift, and time is of the essence, the geostick reading, if previously correlated, might be a valid substitute.

5.7   Percent Compaction Specifications for Clay Fills

In Chapter 4 I presented a method for specifying percentage soil compaction for *structural* fills, which by definition are granular or essentially granular soils (EGS). The method is not applicable to clays, essentially cohesive soils (ECS), and its use is questionable for a gemicoss. It is natural, therefore, to think in terms of developing a distinct (perhaps analogous) method for application to compaction of cohesive soils.

In response to my suggestion, and under my supervision, Zwingle (1981) has proposed such a method. Because the work is preliminary in nature—Zwingle himself calls it a conceptual methodology—I will not attempt to describe his work in detail. I will, however, provide some of its highlights and provide some commentary and suggestions of my own.

1. Analogous to relating percent compaction for structural fills to density, the percent compaction for clays is related to the unconfined compression strength. For rapid loading, the unconfined compression strength is commonly taken as equal to the allowable bearing capacity for clays.

2. A mold of modified dimensions is used to permit valid unconfined compression tests on extruded, compacted samples. This is necessary because of a length–diameter ratio requirement not met by the original Proctor mold.

3. The liquid limit is postulated as a basis for correlating the required unconfined compression strengths with water contents and, in turn, with compacted dry densities of modified Proctor curves.

4. The important differences in engineering properties of clays compacted wet and dry of optimum, at *equal* densities, is considered. These behavioral differences are largely attributed by most authorities to differences in particle orientation, with the wet side exhibiting a more parallel orientation than the dry side. As a result, clays compacted on the wet side are more highly anisotropic, particularly with respect to shear strength and drainage. Clays compacted to high consistencies (stiff, hard) on the dry side may exhibit detrimental expansiveness when subsequently wetted.

As may be inferred by the foregoing, the compaction characteristics of clays is inherently much more complex than that of essentially granular soils. I would, therefore, recommend that geotechnical specialists be consulted when dealing with such soils. I would especially caution nonspecialists to refrain from making judgments on projects dealing with clay fills where the consequences of failure are severe, notably with earth dams.

5.8 Summary

In this chapter, I have presented descriptions of work that I have been involved with over about 25 years. The range of completeness extends from published work and patents to suggestions and ideas for further work. References on artificial fills are cited within the chapter. Some other references and data have not been published, but I offer to share them with serious investigators who wish to pursue the work further.

5.9 Glossary

(+200)(−200). This nomenclature refers to sieve numbers. The number on a sieve (as 200) refers to the number of openings per square centimeter. Thus, the larger the number, the finer the sieve. The 200 sieve is the approximate separator between fine sand and coarse silt. Thus anything passing the 200 sieve (−200) is fines—either silt or clay or a mixture of both.

Bentonite. A clay of very high plasticity of the mineral family montmorillonite; plasticity index is about 400. In construction parlance, it is called "drilling mud," or slurry, and is used often for preventing the walls of excavations (drill holes, trenches) from collapsing. Mixed with water (in large amounts because of such high plasticity), it forms a viscous fluid that exerts lateral pressure on vertical walls.

Uplift pressure. When water flows under a structure such as a dam, upward pressures develop. These uplift pressures must be considered in design.

Excess porewater pressure. When water is confined in the small pores of a relatively impervious soil such as clay, and an external load is applied, the stress induced by the external load is initially "accepted" by the water in the pores. If one were to measure the water pressure at a point (say) 20 ft below the existing water table, the pressure would be in excess of hydrostatic (i.e., greater than $20 \, \gamma_w$). The amount in excess of *hydrostatic* pressure is termed the excess pore water presssure. As the water drains, or is "squeezed out," of the soil–water system, under the influence of the excess porewater pressure, the pressure would diminish, eventually reaching that of the hydrostatic condition. The process of water being squeezed out of the system in such a manner is called consolidation. (See Section 6.2.6 for further explanation.)

Preconsolidation (sand drains, wicks).   A foundation treatment for inducing consolidation before construction (hence preconsolidation), as described above. The external load is a surcharge, usually a fill at the surface, and the sand drains or wicks are designed and installed to accelerate the process to allow construction to commence after a reasonable time period (usually a year). See also Section 9.1.

Structural slab.   A concrete floor that is designed with enough reinforcing steel to span supporting columns without distress or unacceptable deflection. Slab-on-grade construction means the floor requires support of a fill. *Structural* beams and *grade* beams have the same connotation.

Anchor bolts.   Bolts that are used to pin potentially weak soil or rock to an area (below or within) that is judged to be stable; the bolts are placed or driven into the stable area and there anchored with injected grout (thin cement) or special epoxies (a more recent technique).

6

Fills and Fill Compaction

As described in Chapter 3, there are a number of categories of natural soils that are used as fills, with differentiations based principally on texture (grain size distribution for cohesionless soils, and plasticity (P.I.) for cohesive soils). Of increasing importance are fills composed of waste products and artificial materials (Chapter 5). Each fill type requires different methods of field operation.

6.1 Strength, Stability, and Imperviousness: Contrasting Requirements

A helpful way to consider the contrasts among fill types is to consider the components and functions of an earth dam (Figure 6.1). There are two requirements of the dam: It must be strong enough to impound the water in the reservoir, and it must do so without unacceptable seepage losses.

The most permanently stable soils are those that have a granular (cohesionless) texture, are dense, and free-draining.

Marginally draining soils, such as fine sands and silts, are susceptible to instability associated with seepage and capillarity.

Clay soils, cohesive and plastic in nature, exhibit potentially damaging changes. When these soils are soft, consolidation can occur, leading to perhaps intolerable settlements. When they are hard or overcompacted, the same soils may expand when wetted.

The simplest conception of the design and construction of an earth dam is to provide strength and stability using granular soil for the shells and ensure imperviousness with the clay core wall (see Figure 6.1). A cutoff wall (sheet pile or grout curtain, typically) may be necessary to reduce seepage under the dam to an acceptable level.

I further suggest that the suitability of fills—indeed, all soils—should be considered on the basis of function rather than whether they are good or

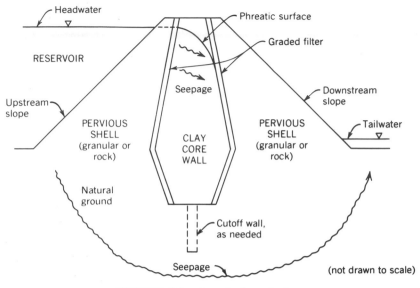

FIGURE 6.1   A typical earth dam.

bad. Clearly, clay is a good fill in the core wall of an earth dam, but bad as a backfill behind a retaining wall.

6.2   Potential Problems with Earth Structures

A second reason for focusing on fills in earth dams in this chapter is the current National Dam Inspection Program, initiated principally because of the major dam failures of the Teton Dam in Idaho and the Taccoa Dam in Georgia. It is probable that this inspection (and remediation) program will require decades of attention. It is also sensible that the construction of new earth dams should be done with careful attention to proper methods of compaction and familiarity with the consequences of poor design and construction will be helpful. To some degree, the material presented in the following sections will also be helpful to those involved in the ongoing inspection of existing dams.

6.2.1   Marginally Permeable Soils

Marginally permeable soils include fine sands and silts, with permeability coefficients in the 10^{-3}–10^{-4} cm/sec range. The problems associated with such soils are potential quick conditions *(boiling)*, liquefaction, and frost heave. With respect to earth dams, the problem of boiling is especially paramount; the failures of the Teton Dam and the Taccoa Dam are both judged to have been caused by the internal deterioration of the dam, an insidious process typically starting slowly with the phenomenon of boiling, with subsequent progression to accelerating deterioration: piping, caving, roofing, and eventual breaching (a sudden, complete collapse).

The Flow Net

Figure 6.2 shows a *flow net,* representing the flow under a sheet pile caused by the impoundment of 12 ft of water. The flow net is composed of intersecting sets of *flow lines* and *equipotential lines.* A flow line is straightforwardly defined as any path traversed by water flowing from headwater to tailwater, thus *ABCD* and *EFGHIJK* in the figure; *LM* is also a flow line, a boundary flow line (as is *ABCD*). Lines 1–2, 3–4, 5–6, 7–8, 9–10, 11–12, and 13–14 are equipotential lines and may be defined graphically as demonstrated by the piezometers shown in the figure: At any point on line 3–4 (*N* and *Q,* shown) the energy of the water may be represented by its position (*Z*) with respect to the reference, and its pressure head (*p*) as represented by its rise in the piezometer tube. Thus, a particle of water at *E* will have lost 2 ft of head in flowing from point *E* to point *F.* This would represent an *equipotential drop* (ep drop). The same particle of water, traversing the flow line *EFGHIJK* would suffer six ep drops, losing 2 ft of head across each drop (by frictional resistance to flow), for a total head loss of 12 ft (to the tailwater). With respect to fluid mechanics, note that the velocity head, $v^2/2g$, is not a factor in flow through soils, inasmuch as the velocity is negligibly small, laminar, and governed by Darcy's law.

The principal requirement of any flow net are (1) that all flow lines and equipotential lines intersect at right angles, and (2) that all elements are

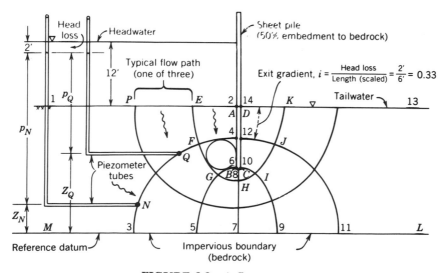

FIGURE 6.2   A flow net.

square. An element is defined as being bounded on two opposite sides by flow lines and equipotential lines (e.g., *FG* 46, Figure 6.2). Since all lines in the flow net are curved, a better definition of square is that a circle may be inscribed in each element of the flow net, as shown.

A *flow path* is defined as the region between two adjacent flow lines *(AE, EP,* and *PM,* Figure 6.2).

Thus, Figure 6.2 exhibits *six* equipotential drops and *three* flow paths.

Partial ep drops and/or partial flow paths are valid, but in such a case all elements within the "partial" must be similar rectangles rather than squares.

The ratio of the number of flow paths, N_f, to the number of ep drops, N_e, is termed the *shape factor* for the flow net. One of the questions that commonly occurs to those learning to draw flow nets is, "How many flow lines (paths) should I draw?" There is no correct answer, as it depends on the particular case, and experience is needed to decide. However, I suggest that most flow nets should contain between two and, perhaps, as many as seven flow paths. It has been my observation that most beginners attempt too many flow paths for a given problem. Irrespective of the number of flow lines selected, however, you should note that there is only one correct shape factor. Thus, in Figure 6.2, if one chose to draw six flow paths, this would, of necessity, result in 12 ep drops. The shape factor would be ½ in either case.

An assist to the beginner in drawing flow nets is to study flow nets correctly drawn by others. There are a surprising number of rather badly drawn flow nets appearing in published books. See how many you can spot! Remember: ep's and flow lines must be orthogonal *everywhere,* and all elements must be square (or similar rectangles in partials).

A feature of seepage analysis and design which is not generally known is that the seepage *quantity* one computes from a flow net is rather insensitive to the quality of the flow net. Thus, one can get a pretty good answer for discharge even with a quickly sketched, "poor" flow net. However, to get acceptable answers for seepage *pressures,* the flow net must be correct, or at least more nearly so than the usual 15-min sketched variety usually produces. This information is valuable, in that much time can be wasted on producing a correct flow net, only to discover that the discharge is unacceptably high. I recommend that one sketch a quick flow net, and check for discharge. When the proposed design is acceptable from that standpoint, the flow net can then be redrawn carefully for purposes of pressure and stability analyses. For example, in Figure 6.2, if the discharge had proven to be unacceptably high for the 50% cutoff depth shown, a second trial at (say) 75% cutoff could be tried, for which, of course, there would result a different flow net. This approach could save days of wasted effort, since drawing a good flow net can often take many hours, particularly for a beginner.

Seepage Loss.   The seepage loss (discharge) is given by

$$Q = k \frac{N_f}{N_e} HL$$

where k is the permeability coefficient (velocity units), N_f is the number of flow paths, N_e is the number of ep drops, H is the total head loss, headwater to tailwater (12 ft in Figure 6.2), and L is the appropriate dimension perpendicular to the plane of the flow net (e.g., length of dam or sheet pile).

Seepage Pressures.   When water seeps through a soil, a seepage force develops in the same direction as the flow. The unit seepage force or seepage pressure is

$$j = i \, \gamma_w$$

where i is the hydraulic gradient and γ_w is the unit weight of water (62.4 lb/ft^3 or 1 g/cc). The hydraulic gradient is defined as *the head loss divided by the length over which it is lost*. Since seepage studies involve the determination of hydraulic gradients at points anywhere within the flow, as defined by the flow net element at the region of the point, I have found that people make fewer errors if they remember the word definition of the hydraulic gradient rather than reducing the definition to a formula. Thus, the hydraulic gradient at the center of element *FG* 46, Figure 6.2, is 2/6 = 0.33 (i.e., 2-ft head loss, divided by 6 ft, the length over which it is lost) (as scaled from the flow net element).

Of particular importance in seepage studies is the *exit hydraulic gradient*, as compared to the *critical exit hydraulic gradient*, i_c. The exit hydraulic gradient is the one associated with the smallest square at exit (tailwater), since this would represent the maximum exit gradient (smallest length over which it is lost). This gradient in Figure 6.2 is 2/6 = 0.33. The critical exit hydraulic gradient is determined by considering the force balance at exit. The seepage force j would be upward, corresponding to the direction of flow, and the resisting downward force would be the effective weight of the (submerged) soil, γ'. The critical exit hydraulic gradient would ocur when the seepage force equals the effective unit weight, or

$$i_c \, \gamma_w = \gamma'$$

Thus,

$$i_c = \frac{\gamma'}{\gamma_w}$$

Since the submerged, effective unit weight of soils would be approximately 60 lb/ft^3 (in accordance with Archimede's principle), the critical exit hydraulic gradient is approximately 1. If such a condition exists through poor design (or no design at all in the case of many old dams), boiling will develop. (In modern seepage design, the facility is designed so that the exit hydraulic gradient does not exceed 0.8, thus incorporating some measure of a factor of safety.) Boiling is the phenomenon that is misleadingly called quicksand in old Tarzan movies, and in many lay publications. It should be apparent from the foregoing, however, that the instability is a condition rather than a material. If the condition is not permitted to develop, the same "quicksand" can be quite stable as a foundation material. The general approach to preventing this condition (i.e., reducing the exit gradient to 0.8 or less) is to force the water to flow through a longer distance by, for example, installing or deepening a cutoff wall.

In the initial stages of boiling, individual particles of sand will be dislodged upward by the seepage forces, and be deposited in a more-or-less symmetrical pile that very much resembles an ant hill. Allowed to continue, the next stage of deterioration will be *piping*, the backward progression of a small tunnel (pipe), starting at the tailwater and progressing backward to the area under the structure (see Figure 6.3). As the piping becomes more pronounced, the flow will concentrate in the small area of the pipe, and, in accordance with the continuity principle of steady-state flow, the velocity will increase proportionally ($V_1 A_1 = V_2 A_2$). Since V is proportional to i, the hydraulic gradient, and $j = i\gamma_w$, the seepage forces will increase

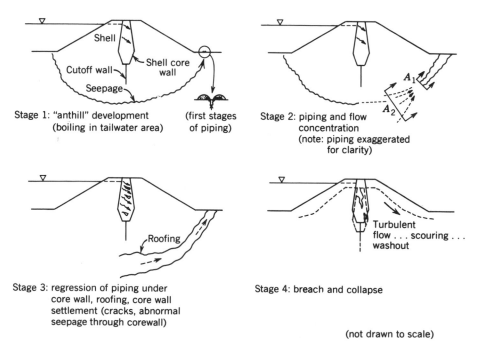

Stage 1: "anthill" development   (first stages
          (boiling in tailwater area)   of piping)

Stage 2: piping and flow
          concentration
          (note: piping exaggerated
          for clarity)

Stage 3: regression of piping under
          core wall, roofing, core wall
          settlement (cracks, abnormal
          seepage through corewall)

Stage 4: breach and collapse

(not drawn to scale)

FIGURE 6.3   Deterioration of an earth structure.

correspondingly. It may be said of piping, "the worse it gets, the worse it gets." In time, the piping will migrate under the structure, grow in size, and eventually *caving* and *roofing* (roof collapse) will occur. Finally, the structure above will settle. In the case of a typical earth dam, Figure 6.1, the clay core wall will develop cracks, and piping will commence *through* the dam, possibly resulting in breaching and complete collapse.

The kind of deterioration and distress just described is judged by experts to have been the cause of the Teton and Taccoa dam failures.

Summary. The foregoing brief descriptions of seepage are included principally to emphasize how important the proper compaction of earth structures, especially dams, can be. For a fuller treatment of all aspects of the subject, Cedergren's *Seepaage, Drainage, and Flow Nets* is highly recommended, particularly for its practical approach to design and construction.

Liquefaction

If an earth mass of marginally permeable soil is loose, saturated, and cohesionless, and is subjected to a shock, the soil mass may instantaneously lose its shear strength and fail suddenly (and often dramatically), a phenomenon called *liquefaction*.

Shocks can be caused by earthquakes, the most dramatic example of which occurred in the 1964 quake in Nigata, Japan, where 6-story buildings sank into the ground or were turned over. An aerial photograph of this dramatic event is shown in Figure 6.4. However, liquefaction can also be induced by less exotic sources of shock, involving everyday, routine construction operations. An excellent example is described by Sowers, an incident involving the abrupt "swallowing" of a bulldozer traversing a loose, saturated hydraulic landfill. (Contractors and dozer operators please note!) Other common sources of shock include blasting and pile driving.

Frost Action

Three factors are necessary for frost action to occur: a frost-susceptible soil, a shallow water table, and (seasonally) cold weather. The phenomenon is caused by the growth of ice lenses that are fed by upward capillary flow from the water taable during periods of cold weather. The combined effect of the ice lenses is to produce frost heave, which is potentially damaging to any structure in or above the soil. The worst damage, however, usually occurs in the early spring. Since thawing will occur initially near the surface, the meltwater will be temporarily trapped by the still-frozen subsoil. During this time period, severe damage can occur to highways by an action called *pumping*, the high-energy ejection of water and soil in a lateral direction caused by the vehicle live loads. Heavy trucks employed for testing purposes at the AASHTO Test Road in Illinois in the early 1960s were responsible for ejecting large quantities of aggregate, some pieces inches in size, outward

FIGURE 6.4   Liquefaction: Nigata earthquake.

to the shoulders of the test road. This action can produce large voids beneath the pavement, eventually leading to collapse of the pavement.

Frost action is of special concern to those involved in route construction, maintenance, and infrastructure repair. With buildings, the solution is rather simple; we just place all exterior footings beneath the frost line, a depth that is known for any given region, and is usually specified in the local building code. For highways, pipelines, and railroads, we do not have so simple a solution. The solutions here are avoidance (of frost-susceptible soils) when possible, or the incorporation of some type of subbase drainage that will prevent capillary rise. The details of such designs are not within the scope of this book.

The most frost-susceptible soils, according to Burmister (1955, p. 144), are coarse silts. This soil size exhibits a substantial capillary rise, combined with a *rate* of capillary rise that is sufficient to foster the rapid growth of the ice lenses during periods of very cold weather.

Closure

As may be inferred by the foregoing descriptions of the potential problems with marginally permeable soils, one does not consciously opt to use such soils in modern earth structures if one has such a choice, or if one can afford not to use them. Thus, for example, in Figure 6.1, the only region in which such soils would occur (ideally) would be beneath the dam: the natural ground. The shells and core wall would be compacted free-draining, granular soils (or rock), and properly compacted clay, respectively. The graded filters would be select fills of a specified granular texture. In the real world, of course, we must often make do with whatever materials are available. With respect to dams, we must also recognize that many existing dams were built before the development of rational approaches to design. Thus, there undoubtedly exist many dams whose composition are nothing more than embankments of dumped material of more or less the same material throughout, but necessarily containing enough fines to be rated somewhere between marginally permeable and impervious. In such a case, there is a danger that the phreatic surface (the top flow line of the flow net, Figure 6.1) will extend to and emerge from, the downstream slope, thus creating instability there, including erosion and eventual slope failure. In modern design the porous material of the shell prevents this flow pattern.

6.2.2   Compacted Clays

Compaction curves presented earlier depicted an approximately symmetrical bell-shaped curve (except for select fills), showing a "dry side" and a "wet side." Thus, at any density level, a horizontal line would intersect the compaction curve at two points such as *D* and *W* in Figure 6.5. Since both soils have the same dry density, it might be inferred that both soils have the same engineering properties. For clays, this is not true, and the differences in properties can be significant and pronounced. The differences are caused primarily by the complex *structure* of clays, in turn caused by the size and shapes of clay particles. As determined by electron microscopes,

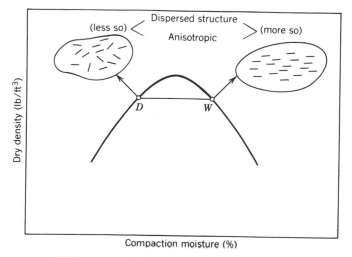

FIGURE 6.5   Properties of compacted clays.

all clay particles are either flat or needle-shaped. Because of their submicroscopic size, the electrochemical forces that develop in their interaction with water are more predominant than gravity (weight) forces. Thus, when a clay is deposited in water, a natural structure develops that includes a complex array of arches among individual particles—a "house of cards" arrangement—termed a honeycomb structure (flocculent in saltwater environments, where the structure is more complex).

When such soils are compacted, there is partial breakdown of the structure toward a more-or-less parallel orientation of the flat (or needle-shaped) particles. This is called a *dispersed* structure. As shown in Figure 6.5 in a schematic way, the tendency toward parallelism is greater on the wet side *(W)* than on the dry side *(D)*. In general, it is this difference in particle orientation that causes differences in behavior.

Lambe (1958) has studied the differences in engineering behavior of pairs of compacted soils suggested by Figure 6.5 *(D* and *W)*. Following is a brief summary of some of his conclusions.

1. Soils compacted wet of optimum *W* are more anisotropic than those compacted to the same dry density on the dry side *(D)*. Anisotropy refers to materials that have different properties in different directions.

2. As a corollary to this, and referring to Figure 6.5, it is evident that soil *W* has higher permeability in the horizontal direction than does soil *D*.

3. Soil *W* has lower shear strength in a horizontal direction than soil *D*.

4. Soil *D* has higher shear strength, in general, than soil *W* in the as-compacted condition, since it has a lower moisture content.

5. However, soil *D* also has a greater potential for expansion upon wetting (a potential negative characteristic).

6. Analogously, soil *W* is weaker, more compressible, and has less potential for expansion than soil *D*.

When writing specifications for the compaction of clays, then, the designer often calls for compaction to be performed 2 or 3% "wet of optimum," or "dry of optimum," depending upon the circumstances and function of the fill. For example, designers commonly call for clayey fills that are to support floor slabs or grade beams to be compacted wet of optimum, because clays compacted on the dry side would have the potential to expand upon subsequent wetting and cause damage to the slab or beam. (Of course, a better design would be to call for a select fill for support, but that is not always economically feasible.) As a contrasting example, if one was considering the compaction of the subgrade of a highway, the height of the embankment would be a factor in the decision. For a high embankment, say 20 ft, dry side compaction of the subgrade would provide the greater strength desired, and the potential expansiveness would not be of significant concern because of the high confining pressure of the 20-ft embankment.

By similar reasoning, in a region of the highway where the embankment thickness is small, the wet side would be suggested for subgrade compaction.

As a final example, the clay core wall of an earth dam would be compacted wet of optimum to avoid the higher horizontal seepage and concomitant piping potential associated with the more pronounced dispersed structure of clays compacted wet of optimum. (See Figure 6.5.) In this case, the function of the clay core wall is seepage control; the necessary strength is provided by the shells of the dam.

These examples illustrate the contrasting requirements of strength, stability, and imperviousness.

6.2.3   Select Fills

As described in Chapter 3 and depicted in Figure 3.4, select fills are free-draining soils, predominantly sands. The term *select fills* is becoming more and more meaningful in that soils of such texture are becoming scarcer and thus more expensive. Because of their cost, specifications and designs are often formulated to minimize the quantity needed. Frequently, specifications will call for a pipeline to be seated on a select fill in a backfilled trench, but the fill above the pipe will be the soil that was excavated, assuming such soil is not of disqualifying texture such as peats or highly plastic clays. Similarly, in the case of an earth dam located in a region where select borrow is unusually scarce, a dam may be designed somewhat as shown in Figure 6.6. Instead of having a pervious downstream shell, requiring large

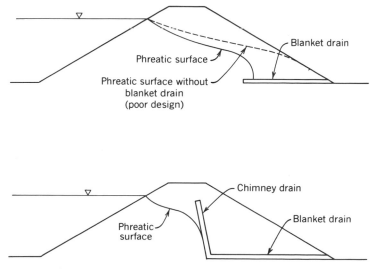

FIGURE 6.6   Minimizing select fill in an earth dam.

quantities of select fill, as in Figure 6.1, an outlet blanket is provided, and the flow (phreatic surface) is thus contained safely within the dam, as shown. In some cases, the blanket is extended to, and even up, the face of the clay core wall: a chimney drain.

Another use of select fill is the *graded* filter shown in Figure 6.1. It is a basic tenet of good seepage design to provide a gradual transition for the flow of water. In Figure 6.1, for example, without the graded filter on the downstream face of the core wall, the clay would be washed out into the large pore spaces of the shell, a form of piping which could cause serious deterioration of the core. Cedergren (1977) provides criteria for selecting the texture of a filter material for the desired transition.

Select fills may also be used as structural fills to support foundation elements. When compacted to high relative densities, such fills provide the most permanently stable foundation, since they are not susceptible to subsequent changes that may afflict clays or marginally draining soils. One caution: Since select fills are cohesionless, they must be laterally contained, even when dense, to retain their stability. Thus, one must be cautious when excavating adjacent to a foundation element supported by a granular soil, lest the removal of lateral support cause a lateral flow of the soil from beneath the foundation. Such an occurrence can be both costly and embarrassing.

6.2.4 Rock Fills

A properly placed rock fill will provide the necessary strength, stability, and perviousness for an earthen dam, as indicated by the upstream and downstream shells of Figure 6.1. This assumes that the void spaces between rock particles (cobbles, boulders, blocks) are not filled with finer soil particles that would inhibit flow.

The principal difficulties encountered with the placement and compaction (if any) of a rock fill are caused by the size of particles, their shape and surface texture, and the geologic soundness of the individual rock particles.

The angularity of rock particles will depend largely upon the nature of the parent rock body from which it was excavated. As extremes, rock obtained from a stream terrace would be composed of rounded, smooth cobbles; a "ripped" sandstone would be "slabby" and of rough surface texture; and a rock obtained from blasting in a sound rock body might be "blocky" and angular.

As mentioned in Chapter 5 (Section 5.1.4), highly angular soils (and rocks), *if dense,* are extremely strong and stable because of the interlocking of particles; the problem with angular rock fills is getting them that way. In the case of a slabby sandstone, there is a danger of very large triangular void spaces being created by bridging. Large void spaces can also be formed in blocky, angular rock of generally uniform sizes. In the latter case, there would be many point contacts within the rock mass. As the height of the fill increases, the stresses at these points would increase to very high levels because of the small contact areas. The danger lies in the postconstruction failure of large numbers of individual rock particles at contact points or edges, resulting in major settlements.

The general approach to preventing this is to induce such failures during construction, either by using very heavy rollers, where feasible, and by keeping the rock wet throughout the construction period. (There is one incident on record where an earth–rock dam was completed "in the dry". When water was subsequently impounded, the upstream rock fill settled 10 ft, undoubtedly caused by the weakening of the rock, by wetting, at the myriad points and edges of high stress concentrations, resulting in localized cracking, slipping, and shear failures, the aggregate effect of which was sudden settlement. This in turn damaged the core wall, and the entire dam failed.)

The wide variety of factors that can influence the postconstruction behavior of a rock fill makes it a difficult task to decide which methods of control should be employed for a given job. Not the least of these is the importance of the fill in the contexts of both cost and consequences of failure. Clearly, a large earth–rock dam would warrant much greater control than a parking area for a warehouse facility in the back country. The following suggestions are possible considerations:

1. The evaluation of the proposed rock fill, from the standpoint of geologic soundness of the eventual (excavated) rock particles should be made by those with expertise in this area: a geotechnical engineer with experience in the growing discipline of rock mechanics, or an engineering geologist. Rock weathering, geologic defects, mineralogy, and structure of the parent rock will greatly influence fill quality.

2. In large slabby rock fills, it may be sensible to limit the angle of placement repose of the larger slabs to (say) 15 degrees, thus reducing potential postconstruction settlements.

3. For small slabby rock, it may be feasible to use heavily loaded sheepsfoot rollers to fracture the slabs into even smaller pieces. I have seen this work quite well where the parent material was a badly weathered sedimentary rock.

4. As a minimum control, repeated here for emphasis, the rock should be continually wetted as the depth of fill increases, the purpose being to induce failures at point contacts in a gradual, controlled fashion as the height of fill increases.

6.2.5   Silt Fills

Silts (or essentially silty sols) are, in my opinion, the poorest of the four soil types one might use as a fill, in the sense that they add little strength to the soil (as with clay binder), but do impart undesirable marginally draining properties to the soil by clogging the voids of what might otherwise be a free-draining soil mixture of (say) sand and gravel. Thus the soils would be susceptible to instability associated with seepage pressure (boiling, piping), liquefaction, and frost action (Section 6.2.1). In addition, the soil would be very difficult to compact because of its moisture sensitivity (see steep compaction curve in Section 3.3).

6.2.6   Embankment Stability

When an embankment is constructed for whatever purpose, a design choice must be made with respect to the slopes of each side. Clearly, steeper slopes are cheaper in that they require less material and construction effort, and in the case of route construction, less right-of-way. But the slopes must also be shallow enough and strong enough so that they do not fail. In addition, in the case of retaining structures such as earth dams, they must not fail by sliding (at the base of the embankment), or by being overturned. Nor can they be allowed to fail by internal deterioration processes such as have been described in the preceding sections.

Except for extraordinary events such as major earthquakes, freely draining materials such as rock, gravel, and the coarser sands will not fail in the classic, sudden, dramatic sense, as long as they are reasonably well compacted. Distress associated with such materials generally results from intolerable settlements.

Clay soils and marginally permeable soils can and do fail dramatically, including slope failures, bearing-capacity failures, horizontal sliding, and overturning. More often than not, the final failure is a result of long-term internal deterioration (piping) or some other water-related phenomenon caused by the soil's inability to drain freely and quickly. Actually, slope failures, bearing capacity failures, and horizontal sliding are similar in that they all involve *shear* failure, as illustrated by Figure 6.7.

Perhaps the best illustration of how a slope failure is often a water-related phenomenon is to cite what is generally regarded as the most probable occurrence of slope failure with earth dams—the time period following rapid drawdown. There are at least two types of hydraulic structures where the level of the impounded water drops substantially and rapidly on a regular basis, single-purpose flood control dams, and the reservoir embankments of a pumped-storage hydroelectric facility. With respect to the latter, the rapid drawdown generally occurs daily. (Rapid here means a matter of hours.)

What happens is illustrated in Figure 6.7*a*. The water level in the reservoir drops to the lower level. As a result, the weight of the soil, previously submerged and thus effectively having a unit weight of about 60 pcf (Archimedes' principle), assumes an effective unit weight of 122.4 pcf, approximately doubled. However, if the soil is clayey or silty, the porewater

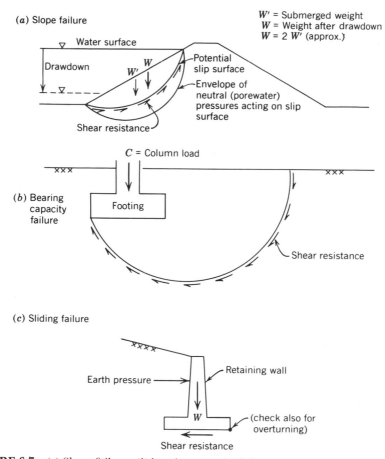

FIGURE 6.7 (*a*) Slope failure; (*b*) bearing capacity failure; and (*c*) horizontal sliding: all shear failure.

pressures existing along the potential (shear) failure plane, do not dissipate for some time after the drawdown because of their slow-draining nature. Since porewater pressures do not contribute to the effective shear resisting capability of the soil, the potential for slope failure is greatest during this time period, inasmuch as the driving forces (the weight of the soil) have doubled (approximately), but there has been no change in the resisting forces (the shear along the potential failure plane), nor will there be any until such time as the pore pressures dissipate as a result of drainage.

6.2.7 Summary

In this section, I have described briefly an array of problems that can and do occur in earth structures, in the belief that an awareness of such problems will enable designers, contractors, and inspectors to deal with compacted fills more effectively.

The interaction of water and soil in marginally permeable soils and the attendant problems of piping, liquefaction, and frost action seem to create most of the problems.

Potential problems with compacted clays, while not as commonly encountered, are nonetheless of equal or greater importance, inasmuch as the failure of earth dams can be and has been catastrophic.

Problems with select fills and rock fills are also outlined, and minimum precautions such as vibratory compaction and continous wetting are recommended. Brief considerations of embankment stability conclude the section.

6.3   Controlled and Uncontrolled Fills

Throughout my experience in professional practice, and in all my reading about fills, the terms *controlled* and *uncontrolled* fills have been used in specifications, conversations, and the literature, implying that a fill must be one or the other. I recommend, however, that an intermediate category is sensible—a *partially* controlled fill.

It is helpful to one's evaluation of a fill (or any soil or rock, for that matter) to consider two factors—texture and condition. (See Chapters 2 and 3 for fuller commentary on this recommendation.) For fills it is often the case that the texture is controlled to some degree, but not the condition (i.e., placement and compaction control). Simple evidence of this is the frequent signs one sees along highways: "Clean Fill Wanted." If such a fill has been in place for many years, its own weight, seepage forces, and frequent trafficking or parking may result in sufficient inadvertent compaction to render the fill quite suitable for light to moderate structural loadings. Alternatively, if the texture is judged acceptable, preferably by test pit inspection, but the condition (density) is not, it may be feasible to compact the soil to acceptable densities by some procedure.

Thus, a fill, once identified as such, should not automatically be rejected for structural loadings just because it is a fill. The important first key is *texture*. Generally speaking, any significant amount of organic material should disqualify the material as a load-bearing fill. Chunks of concrete, or masonry should not be a basis for disqualification—mattresses and garbage, yes!

6.4   Nonstandard and Special Fills

The fill types described heretofore have been either of the standard variety (Section 6.2) or of the new, research-oriented type (waste materials and artificial fills, Sections 5.2 and 5.3, respectively). There are a number of fill types that may be classed as nonstandard or special. These include (1) hydraulic fills, (2) sanitary fills, (3) rock–soil fills, and (4) rubble fills.

6.4.1   Hydraulic Fills

Where circumstances and proximity are favorable, a site may be filled by pumping a soil–water mixture to the site from a nearby body of water, typically from a dredging operation of a river or bay (Sowers, 1979, p. 263). Often these bay muds are composed of predominantly silt sizes. Moreover, the economics of the operation are such that loose lifts are several feet or more in thickness. The ideal procedure would be to allow sufficient time for the fill to settle partially under the influence of drainage pressures and its own increasing weight or induce drainage by installing sumps or even well points (at considerable expense). Further compaction can then be accomplished by either rolling, by vibroflotation, or by dynamic compaction. (Compaction processes will be described more fully in Section 6.5, following.)

6.4.2   Sanitary Fills

Although this book deals with load-bearing fills, some mention should be made of one nonstructural fill because of its increasing importance: sanitary landfills. I will limit my commentary to listing the requirements stipulated by the United States Environmental Protection Agency (EPA), and will point out the observation by Laguros and Robertson (1980) that there often exist problems of conflicting requirements. For example, the most suitable soil for leachate control (the containment of polluted water) is clay but it is the worst soil for venting potentially explosive gases.

According to the EPA (1972), sanitary land filling is an engineered method of disposing of solid wastes on land by spreading them in thin layers, compacting them to the smallest practical volume, and covering them with soil each working day in a manner that protects the environment by:

1. Preventing the entrance of rodents to the compacted refuse (vector control).
2. Preventing flies and odors from emerging.
3. Minimizing the entrance of moisture.
4. Allowing the venting of decomposition gases.
5. Preventing leachate seepage.
6. Inducing the growth of vegetation for final cover.
7. Providing a road base to support vehicular traffic.

6.4.3   Rock–Soil Fills

On many jobs, any rock excavated at the site is designated as waste in the specifications, thus imposing a double cost in that it is not used as fill and there is a cost associated with its disposal.

I suggest that on some jobs it may be feasible and cost-saving to use the rock as fill with alternating lifts of soil. On one job, a rippable sedimentary rock was used in this way, with the added procedure of specifying passes with a heavily loaded sheepsfoot roller over the lifts of slabby rock; the high contact pressures of the roller fractured the slabs of rock and forced the broken pieces into more-or-less horizontal orientation. The subsequent lift of soil (a TALB) was placed with the idea that it would, upon standard compaction, fill the large void spaces of the rock below. This procedure, while unusual, worked quite well and to the satisfaction of all concerned. It is an example of making do with available materials, a procedure that is becoming more and more necessary as sources of better fills become scarcer. (One colleague recently confided to me that he had to "scratch around" from three or four sources to obtain sufficient borrow for one job. This suggests that the problem of changing borrow (Section 4.3) will become more acute.

6.4.4   Rubble Fills

An assortment of nonstandard materials were described in Chapter 5 and were referred to as either waste materials or artificial fills. These were, and have been investigated as, research projects. One additional waste material, while not generally amenable to research investigations, is demolition rubble. I believe that such materials can serve as an additional source of fill if selected and used in accordance with a variety of common-sense guidelines, many of which have already been presented. Thus, materials such as chunks of reinforced concrete, asphalt, bricks, and stone could serve as acceptable fill components. Techniques of placement and compaction could be the same as described for rock–soil fills in the preceding section. Also, one might consider placing acceptable rubble materials in alternate layers with some other waste material, such as fly ash, to facilitate compaction.

The principal disqualifying factor for the fill material should be based upon answers to the following questions: Will the material deteriorate or rot during the projected service life of the fill? Will unacceptably large void spaces be created by its placement? Thus, wood or other organic trash, and metal would be excluded. I would not include reinforcing bars as disqualified, inasmuch as much of their surface area would presumably be encased in concrete. Sheet metal is another matter, in that it could create large void spaces by bridging when placed. After rusting, soil from above could settle into the exposed void spaces, probably resulting in significant surface settlement.

6.5   Compactors and Lift Thicknesses

In Section 3.1 I suggested that the most helpful perception of the compaction process was to view it as the expulsion of air from the soil mass (thud–whoosh!). If one considers that a soil can have an *air* permeability, analogous and probably directly related to the more familiar fluid permeability, the relative lift thicknesses of various soil types necessary for efficient compaction become evident. Table 6.1 depicts the approximate range of lift thicknesses for five soil types shown in Figure 3.4; also shown are suggested compactors.

A great variety of compactors are available for soil compaction, including different variations of the same class (i.e., sheepsfoot rollers). Fletcher and Smoots (1974, p. 291) and Schroeder (1980, p. 133) describe and present photographs of an assortment of compactors.

Sheepsfoot rollers are available with different foot sizes and shapes, probably the most important distinction inasmuch as the contact pressure is directly related to the size of the feet and, of course, the toal weight of the roller, including ballast. (Almost all drum rollers are hollow, so that the total weight can be controlled by the choice of ballast with which to fill the drum—none, water, water plus sand.) When compacting a loose lift of clay, the feet will sink deeply into the soil during the first pass. Upon sucessive passes, the roller will sink less deeply *if proper compaction is being achieved.* The roller is said to "walk out," compacting from the bottom up.

Compactors that are most suitable for silts or silty soils are either the pneumatic rubber-tired rollers, which provide a kneading action, or a roller

TABLE 6.1. Lift Thicknesses and Compactors for Various Soil Types

Soil Types	Lift Thickness (in.)	Suggested Compactor[a]
Fat clays	3–5	Sheepsfoot roller
Lean clays	4–7	Sheepsfoot roller
Silts	5–8	Pneumatic (rubber-tired) or tamping-type roller
TALB	8–15	Pneumatic, wobble-wheel roller; smooth drum roller; or smooth vibratory roller
Select fills	24(+)	Smooth drum vibratory roller

[a]For more detailed and comprehensive advice regarding selection of compactors, several organizations have devised charts and tables that list properties and construction characteristics for soils classified according to the Unified System, a system containing 17 categories of soils.

NOTE: As stated earlier, it is not within the scope of this book to go into details of soil classification. For those unfamiliar with the Unified System, I recommend that you learn the system, or retain a soils specialist to establish the correct classification(s) for the soil(s) of interest. The information may be found in Means and Parcher (1963, p. 119) or Ritter and Paquette (1960, p. 206), or in other standard texts under the approximate heading of Classification of Soils.

of the tamping type, which provides an impact–vibration action. Which type one chooses depends to a large degree on the plasticity of the fines in the soil. Clean silts (those possessing negligible plasticity) would compact best with a vibratory action, but those with significant plasticity (ML in the Unified System) would compact more readily under the action of the kneading of the pneumatic roller. It should be remembered, however, that silts are inherently difficult to compact because of their moisture sensitivity and should be avoided as a choice of fill where such a choice is possible.

TALBs can be compacted efficiently with an assortment of different roller types. Again, the choice depends largely on the plastic nature of the finer soil fraction.

Select fills should always be compacted by heavy *vibratory* rollers, with "excess" moisture (puddling), and in lifts of 24 in. or more. I am advised that so-called supercompactors exist that can compact lifts as thick as 10 ft. Only in hydraulic filling operations would one normally encounter lifts of this thickness, and for emphasis it should be remembered that compaction passes should be made with such rollers only after earlier stage compaction has been achieved by some form of drainage. (See Hydraulic Fills, Section 6.4.1.)

Occasionally one will encounter a natural, thick deposit of granular soil that has excellent texture but is too loose to support even moderate loading. In such a case, it may be advantageous to employ a technique called vibroflotation. According to Fletcher and Smoots (1974, p. 295), thicknesses of 30–40 ft may be compacted. The vibrofloat is essentially a large vibrating probe that is inserted into the ground and vibrated within the hole while simultaneously injecting or jetting water. As a result the soil "flows" toward the probe, creating a cone of depression much like water flowing to a well during pumping. Sand is more or less continuously backfilled into the depressions surrounding the probe. Soil gradation is an important factor in determining feasibility. It is my belief that angularity of the soil may be of equal or greater importance. (See Section 5.4 for additional commentary.)

For important earthwork projects, vibroflotation might also be considered for the final stage in the compaction of a hydraulic fill.

Another technique for stabilizing thick deposits of soil is *dynamic compaction* or *dynamic consolidation,* a procedure involving the dropping of very heavy weights (12–200 tons) from substantial heights (60–120 ft). This is a relatively new technique but one that is gaining acceptance. Unlike vibroflotation, it seems to work in soils of clayey texture as well as in granular soils. Loose or soft deposits up to 50 or 60 ft thick have been stabilized successfully. As may be inferred by the numbers cited, the impacts are substantial, and the resultant vibrations may prohibit dynamic consolidation in urban areas. As with vibroflotation, such exotic work requires the careful evaluation of geotechnical and construction specialists. (Because the focus of this book is primarily ordinary techniques of compaction stabilization and the intended audience is mostly nonspecialists, details of vibroflotation and dynamic consolidation are omitted. For those wishing to study these specialized topics fully, I recommend the Geodex Information Retrieval System. I expect that most geotechnical consulting firms and some engineering libraries would have the system and would provide bibliographic information on either—or any—geotechnical subject, at little or no cost. Alternatively, the system itself could be purchased, but it is rather expensive for all but specialists or libraries.)

6.5.1   *Making Do*

One type of compactor not yet mentioned, principally because it is not normally used for compaction, is the standard Euclid truck, or Euc. On one job where we had to make do in many phases of the job, Eucs were used effectively by loading them with soil and overinflating the tires to about 60–70 psi. (Standard pneumatic rollers go up to about 95 psi.) Although it is a little awkward because of the missed area between the rear tires, the contact width of the four tires was a surprising 8 ft (approximately).

One of my fun discoveries in researching materials dealing with soil compaction was a classic attempt at making do, described in the article "The Uselessness of Elephants in Compacting Fill," by Richard L. Meehan in the *Canadian Geotechnical Journal,* September 1967. One of the more interesting findings of this "modest field investigation" conducted in Thailand, was that an elephant quickly learns to retrace his (her?) steps, avoiding the softer uncompacted areas by exploring the terrain ahead with its trunk (a remarkable sensory organ, according to Meehan). Thus, uniformity of coverage cannot be accomplished readily. I guess one can also conclude from this that elephants are smarter than sheep—which I always suspected.

6.5.2   *Confined Areas*

I recommend the following experiment to all who read this book. Take a supply of sand, dry it in an oven, and pour it into a 1000-cc graduated cylinder, using a miniature tremie fashioned from a plastic cone and a fitted plastic tube. Fill the cylinder without free-fall (to simulate dumping a fill) to somewhere around the 950-cc level, and record the level. Optionally, place a cylindrical weight on the soil to provide a static confining pressure, and note that the volume change will be negligibly small—even under substantial static pressures. Now strike the cylinder repeatedly with a hammer for about one minute. You will observe that the volume decrease will be about 10%.

This simple, inexpensive experiment illuminates the most problematic of all filling operations: the all-too-common lack (or elimination) of compaction in confined areas, most notably trenches and behind basement walls and retaining walls. For a variety of reasons, it is perhaps understandable that this is often the case. Compaction in confined areas is inherently a labor-intensive operation, and we live in a society where unit labor costs are usually high. In addition, proper compaction is often time-consuming, so the total costs increase accordingly. Finally, because the amount of backfill is usually small, backfilling without proper compaction can usually be done rapidly.

However innocent the practice of dumping backfills may appear, the consequences can be severe. The following case history from my consulting experience will illustrate what I mean.

A Trench Backfill

A lateral utility trench on a highway was alleged by an attorney to have created a traffic hazard and caused his client to lose control of his car, resulting in a crash and serious personal injury. I was retained to review the situation, prepare a report of my findings, and to provide expert testimony if needed. Investigation revealed that the 7-ft-deep trench had been backfilled by pushing the fill into the trench with a front-end loader, and compaction was achieved "with passes of the wheel of the loader" (a quotation from the contractor's own notebook). The fill "was of a granular texture." With this minimal, but very meaningful, information established, the computation of an approximate maximum potential settlement was made, as follows:

1. Assume that the relative density D_R of the dumped fill is 40%.
2. Assume that the maximum and minimum values of the void ratios for the granular fill are 0.8 and 0.5, respectively.
3. Assert that the documented method of compaction had no significant effect on the densification of the fill (i.e., was totally inadequate).
4. Assume that the settlement is given by

$$\Delta H = \frac{H\ \Delta e}{1 + e_0} \tag{1}$$

where Δe is the change in void ratio, e_0 is the original void ratio, and H is the thickness of the fill. The relative density is

$$D_R = \frac{e_{max} - e_0}{e_{max} - e_{min}} \tag{2}$$

Substituting numerical values and computing e_0 (in this case, the dumped void ratio)

$$0.4 = \frac{0.8 - e_0}{0.8 - 0.5}$$

$$e_0 = 0.68$$

Thus, $\Delta = 0.68 - 0.50 = 0.18$. That is, the dumped soil of $e_o = 0.68$ would have its void ratio reduced to (possibly) its minimum value $e_{min} = 0.5$, mostly by the influence of traffic vibrations and impact. Computing the settlement,

$$\Delta H = \frac{(84)\,(0.18)}{1.68}$$

$$= 9.0 \text{ in.}$$

Such a settlement could occur (given enough time) in the trench beneath the pavement. With such a deep void space, it is likely that the pavement would collapse into the hole created, and a serious traffic hazard would indeed result.

Observe the correlation between the little experiment and the computations; thus the 10% figure is a nice round number to keep in mind when estimating the potential settlement of a dumped, granular fill. (Note, finally, that the analysis would not pertain to nongranular fills inasmuch as the range of void ratios for clays and clayey soils is usually much greater. Also, clay fills will settle more readily under the influence of static loads rather than vibration. Shocks, however, can cause flocculent or honeycombed clays to collapse very suddenly.)

Retaining Wall Backfills

A second important type of confined area is behind retaining walls, including basement walls of structures. Ralph Peck and H. O. Ireland (1957) published a paper entitled "Backfill Guide" in which they state:

It is unfortunate that the problems of working in confined space should occur where the need for good fill is the greatest. The cramped working space, the relatively small volume of fill involved, and the backfill material locally available are undoubtedly responsible for much lax enforcement of specifications.

This repeats and augments some of what I have said. Peck and Ireland also present some very instructive and revealing calculations for lateral earth pressures that clearly illustrate the importance of both fill texture and compaction. For dense sand, loose sand, and soft wet clay (or silt), the pressures behind a 15-ft retaining wall were, respectively, 1.69, 2.70, and 5.62 tons/lin ft; for a 30-ft wall, 6.75, 10.0, and 22.5 tons/lin ft. From these data, it is evident that good texture and proper conpaction (plus good drainage) have a strong influence on design of subsurface structures and specifications for selection and placement of fills. Enforcement of specifications is no less important.

A few additional observations:

1. If you apply the 10% criterion suggested by my experiment and supporting computations to the 15 ft of loose sand, it does not bode well for an outdoor patio.

2. If a wet clay is used as backfill for a basement wall—as all too often happens—the following scenario may (and did) occur. (This is one of my favorite case histories, as related in a lecture by the late Jacob Feld.) After placement, there was a period of very dry weather, causing the clay to dry and shrink away from the wall. Thereupon a late afternoon shower occurred, followed by a sudden temperature drop. The accumulated rainwater in the space between the clay and the wall froze, exerting pressures on the wall sufficient to push it over. This simple case history, in addition to being a nice construction version of Murphy's law,* graphically illustrates two important facts: the poor quality of clay as a structural fill and the importance of drainage. Not only should a fill be free-draining itself, but water should not be allowed to accumulate behind the wall. In the case of a basement wall, a drainage system must be incorporated in the design that will divert water around or away from the wall. For a retaining wall, suitably spaced weep holes are placed to allow water to flow through the wall.

Compactors and Compaction Techniques

It is evident that the compactors listed in Table 6.1 cannot generally be used for compaction in trenches and behind walls. Instead, one must use hand-operated tampers, most of which are activated by air compressors or internal combustion engines. Fletcher and Smoots (1974, p. 294) and Sowers (1979, p. 254) describe common tampers. As would be expected, the lift thicknesses would be necessarily smaller than those suggested in Table 6.1, generally of the order of 3–6 in. Unfortunately, there appears to be little information available on how much tamping is required for each layer.

Similarly, there is a dearth of information on compaction moisture. As emphasized in Chapter 3, the optimum moisture content for a soil is a function of the energy of compaction. It follows that if we don't know what energy we are using, we cannot stipulate moisture levels with any degree of specificity. One thing that should *not* be done, however, is the practice of jetting or flooding in attempting to compact soils in confined areas. At first reading, the statement may seem contradictory to what I stated in Section 3.3.3 where puddling was generally recommended for select fills. The key distinction here is that we are now referring to compaction techniques in confined areas, more than likely confined in such a way that water cannot readily drain away because the boundaries of the confined area prevent such drainage. Jetting, flooding, puddling, or whatever you call it will result in the entrapment of water by the fill boundaries (select fills in trenches, for example) or by the fill itself, (clays, silts), resulting in inefficient compaction because of submergence, weakening of the resultant fill, and

*If something can go wrong, it will.

the potential increase of lateral pressures on walls. I like V. J. Brown's characterization (1967, p. 69):

[The entrapment of water] would be the same as if we had a huge rubber balloon filled with water beneath the backfill. A vertical load placed on that balloon will transmit equal forces in all directions.

Brown also states, and I concur,

Too many 20-year and 30-year old specifications that allow flooding still govern trench backfill work, particularly for municipal and for subdivision construction.

The key to whether puddling will aid compaction will be an affirmative answer to the following two questions: (1) Is the soil free-draining? (2) Will the water readily drain through and away from the fill area? Generally, this would apply only to an "open" embankment of select fill. In such a case, the combined effect of puddling and heavy vibratory rolling produces the desired compaction result.

Closure

From what I have described in this section, it is evident that proper compaction in confined areas is a subject that has been badly neglected. To the extent that I have been able to discover, very little research has been done. I expect that this is true because full-scale field testing would be required, requiring expensive excavation machinery and a broad assortment of tampers. The time investment for a well-controlled project would also entail additional major costs.

The one research project of which I am aware is described by Brown in "Soil Compaction in Narrow Places" where one type of impact compaction was investigated. The Arrow hammer was used for the study. The hammer, with its compaction foot, weighs about 1350 lb and incorporates up to a 9-ft drop. This is clearly not a hand tamper! And the projects for which it would be suited are much larger than the trenches, basement walls, etc. that I have in mind—those tens upon thousands of smaller jobs that would benefit from what we could learn from a carefully controlled research project. I close this section by suggesting that such a project could be funded by a consortium of manufacturers of hand-operated tampers. The work could be done at a large university by a team of researchers under the supervision of a senior faculty member with extensive appropriate construction experience.

Finally, I suggest consideration of the use of artificial fills in confined areas as described in Section 5.3 and illustrated typically by Figures 5.1, 5.3, and 5.8. A specific type of artificial fill of which I have become aware is called Poleset, a pourable, quick-setting, plastic foam. It is currently being considered for a weight-credit backfill application in New Jersey. The unit cost of such fill materials is quite high, so it would not routinely be used in place of soil backfill, but where weight credit is a necessity, the design choice may be viable.

6.6   *Energy and Moisture Control*

As has been emphasized, the approach to controlling field compaction of soils is to test those soils in the laboratory in such a way as to simulate field conditions to the extent that is feasible, technically and economically. Sometimes we violate this for the sake of simplicity. Certainly the impact of a 10-lb hammer on a soil contained in a rigid steel mold hardly simulates the effect of (say) a pneumatic roller passing repeatedly over a soil, but it might be said that we compromise by searching for a "bottom-line" simulation, irrespective of methods. That is, if the density achieved in the field matches that of the specified target value obtained from the laboratory test, we have simulation.

The combination of factors in the field that will achieve our ends is almost limitless, and it would be difficult if not impossible to state which combination is best. But we can observe some guidelines that lead us to a reasonably workable selection. Table 6.1 lists suggested compactor types and lift thicknesses for various soil types. Once a particular compactor type is chosen, its total weight can be adjusted by choosing the ballast. If we wish to compact to very high densities—say 98% modified Proctor—we would choose the largest compactor type and load up with, for example, sand and water, plus metal ballast. We might also favor lower ranges of lift thicknesses and a larger number of passes. On the other hand, lighter compaction requirements for "unimportant" fills calling for 95% standard Proctor might be achieved by a lighter roller with no ballast, thicker lifts, and fewer passes.

Compaction moisture levels depend upon the energy of compaction; the higher the energy, the less the moisture (see Figure 3.4). The addition of moisture in the field, when necessary, is usually effected either by a truck with a sprinkler bar or by the compactor itself, some of which are equipped with this capability. For compaction to be effective, it is often necessary to allow time for the water to soak into the loose lift of fill, or physically to mix the added water with a grader or bulldozer. Figure 6.8 illustrates a typical computation for determining the desired optimum. If the field humidity is very low, and especially if mixing is thought necessary, it is advisable to overshoot in the computation to allow for evaporation. A 3% overshoot is suggested, but field checking is recommended if deemed important enough.

If the borrow is too wet, the soil may be dried out somewhat by aeration by scarification, again using a grader or bulldozer. I discuss various other alternatives more fully in Chapter 8, "Fill Control Procedures."

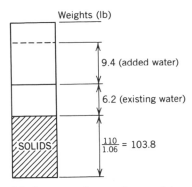

Weights (lb)

9.4 (added water)

6.2 (existing water)

$\frac{110}{1.06} = 103.8$

FIGURE 6.8 Computing added compaction moisture. A borrow has a unit weight of 110 lb/ft³ with a moisture conntent of 6%. Compute the amount of water that should be added to raise its field compaction moisutre content to its optimum of 12%. Assume 3% allowance for mixing loss.

Desired moisture content, w = 15%.
Weight of water, W_w = 0.15 × 103.8 = 15.6 lb
$$\begin{array}{r} -6.2 \\ \hline \text{Added water} = \quad 9.4 \text{ lb} \end{array}$$

$$\frac{9.4 \text{ lb}}{\text{ft}^3} \times \frac{27 \text{ ft}^3}{\text{yd}} = \frac{254 \text{ lb}}{\text{yd}}$$

or

$$\frac{254 \text{ lb}}{\text{yd}} \times \frac{\text{gal}}{8.34 \text{ lb}} = \frac{30.5 \text{ gal}}{\text{yd}}$$

6.7 Glossary

Capillarity. The movement of water (usually thought of as upward movement) in finer-grained soils. Such movement is a major contributing factor to damage caused by frost heave.

Sheet pile. Pile shapes that are driven and interconnected in such a manner as to create a subsurface wall, in seepage design for the purpose of forcing water to flow under the impervious wall thus created.

Grout curtain. Same purpose as above, except here the wall (or curtain) is formed by injecting grout (a thin cement) into porous areas, thus (hopefully) sealing the area and diverting the seepage, as with a sheet pile.

Piezometer. For purposes of this chapter, it may be thought of as simply a hollow tube inserted in the soil to specified points for the purpose of monitoring pressure head (e.g., p_Q, Figure 6.2).

Laminar flow.   Flow of very low velocity, such that Darcy's law pertains. (Except for most unusual cases of rock fissures or "open gravel," laminar flow would be the case in flow through soil or rock.)

Darcy's law.   The velocity is proportional to the hydraulic gradient; the constant of proportionality is called the coefficient of permeability. Thus, $v = k\,i$. (See Section 6.2.1.)

Flood control dams (single purpose).   While many, or perhaps most, dams are multiple-purpose (flood control, water supply, hydroelectric, recreation), if a dam is a single-purpose, flood control dam, its only purpose is to temporarily detain flood waters to protect downstream areas during and shortly after major storms. In such a case, the filling and emptying of the dam is common; hence, rapid drawdown is usual.

Pumped storage.   A hydroelectric generating installation using two closely spaced water bodies at appropriately different elevations for the purpose of generating electricity. This is done by pumping water from the lower body to the upper body during off-peak (overnight) hours, so that the water will be available at the upper location for peak (daytime) demand requirements. The two bodies are connected by a large-diameter conduit called a *penstock,* with turbines located at the lower water body. The typical daily exchange of water creates daily rapid drawdown. The water bodies may be natural or man-made (reservoirs) or a combination of both. The laws of supply and demand make this relatively new system economically viable.

7

Compaction Specifications

Soil Types; Structures Types.   These appellations originated in the early 1960s through some good-natured banter that I engaged in with an academic colleague—a structures type—who, like me, was oriented more toward professional practice than university research. I used to get on him about the occasional phone calls I'd get from structural engineers that would usually be limited to a statement and a question: "I've got this sandy soil, maybe a little gravel. . . . What's the bearing capacity?" (I was always tempted to suggest that they call Karl Terzaghi.) My colleague, who taught structures to the same students to whom I taught soils, was fond of retorting with: "Wow, it must be neat [we talked like that in those days] to be able to come to class with just a pocket penetrometer!"

As I emphasized in the opening chapters of this book, the focus is on practicing nonspecialists, but includes students and young engineers who may become soils specialists or contruction technologists. It is not my intent to presume to advise specialists on how to write specifications; they already know. That being the case, I will concentrate more on how to evaluate specifications others have written, what to look for, what to watch out for. In the process, the potential specialist will also learn to write his or her own.

In Chapter 4 I described some case histories to illustrate major problems in fill control, including those caused by nonspecialists (Section 4.4.1). Thus, the structural engineer who wants to know the bearing capacity is very likely to also dabble in writing compaction specifications.

Morris (1961) has asserted that:

There is a way to write a specification so that the owner will have satisfaction at a reasonable price and the contractor can complete his work with pride and still make a reasonable profit.

I could not agree more.

7.1   Typical Specifications

Studies of earthwork construction practice have shown that most compaction specifications are of either the end-result or methods type, with variations and combinations of both not uncommon.

End-result (or performance) specifications stipulate the results to be achieved, usually a target value density as governed by some percentage of a laboratory compaction test. Other end-result criteria include relative density or some type of plate bearing or field penetration test such as the CBR (California Bearing Ratio). Some percentage of relative density is sensible for select fills, inasmuch as a discrete compaction curve (of the Proctor type) does not exist for such soils. (See Figure 3.4.) CBR values are normally limited to road bases and subgrades, especially and logically where the controlling agency uses CBR values for designing pavement components. Specifications of the latter type, however, are inherently more expensive, both for the laboratory testing and field verification procedures which are required. The more typical target density requirement may be thought of as an *indicator* of CBR values. This is yet another illustration of the notion of the design flowchart in Figure 1.2.

Methods specifications mean just that. They spell out, sometimes rather precisely, the methods and equipment that the contractor must use, including lift thicknesses, field moisture content, and number of passes.

Sometimes method and end results are specified, a practice that is extremely restrictive on all parties, most notably the contractor. In my opinion, this type of specification is also quite illogical because it is naive to think that a rigid set of methods will always produce a desired end result. When they do not, at the very least it can be embarrassing to the writer of the specifications. Moreover, it will frequently lead to arguments and delays in the field, endless change orders, and often legal entanglements—not to mention a lousy job!

A fourth approach, called the suggested-method and end result (Morris, 1961)

allows the more experienced contractor the latitude to make use of his experience, while it offers a guide to the less knowledgeable contractor. At the same time it insures for the owning agency the desired finished product.

Historically, end-result specifications have been more favored by controlling agencies. In Morris's 1961 survey of the United States, only one state had no requirement for percent density. However, fully 80% also specify "in considerable detail the type of equipment to be used in getting the required density."

In general, it appears that the necessary flexibility is afforded the contractor by the avoidance of imposing other requirements such as lift thicknesses and number of passes. In another more recent international survey (Reichert, 1980), 18 countries reported end-result type, 5 had method types, and 10 had requirements of the mixed type. As stated in Morris's earlier study, the 50 states of the United States each had their own requirements, but the majority combined end-result with some form of methods control. In all cases (internationally),

the contractor is required to take the necessary precautions for preventing the materials employed from becoming water-logged during bad weather. The arsenal of conventional methods includes the provision of slopes for water drainage, immediate compaction of soils after spreading, and surface smoothing of compacted areas.

Readers should keep in mind that the two studies that I have cited here dealt exclusively with roadway embankment compaction. Thus, anyone referring to these sources for guidance in preparing or evaluating specifications should be aware that they do not pertain without perhaps important modifications to structural fills, for example, the support of footings. In general, but of course depending upon the specific loading intensities and other circumstances, the end-result densities one would require would be significantly higher. One should also recognize that there are major differences in the required supporting capacities for roads. Thus, the compaction requirements of the upper few feet of an interstate highway would be more stringent than for the footings of lightly loaded warehouses. Conversely, it is helpful to contrast in a general way the requirements for lightly traveled secondary roads and footing support of heavy industrial buildings or bridge abutments. Thus, the terms structural fill, load-bearing fill, and nonload-bearing have relative meaning only.

One excellent approach to the suggested method and end-result specifications is illustrated by the policy of providing a written suggested guide specification for excavation and placement of compacted earth fill on jobs where such work is a peripheral part of the overall construction. This was the standard procedure of the geotechnical consulting firm where I learned the ropes, Woodward–Clyde–Sherard and Associates. This suggested guide would be a standard document included in all reports to clients, often as an appendix. On jobs where the major focus on the work was to be fill placement, a unique set of specifications would be prepared.

7.1.1 Specification Components

In writing or evaluating specifications the following items should be considered.

1. Stripping and grubbing.
2. Excavating.
3. Acceptable fill texture.
4. Selection of representative samples (for laboratory testing).
5. Fill placement (loose lift thickness).

6. Lift preparation (mixing, moisture adjustments).
7. Compaction equipment (equipment, moisture content, number of passes).
8. End-result stipulation (usually target value density).
9. Weather restrictions and preventive measures pertaining thereto.
10. Control (special criteria, procedures, personnel).
11. Correction.

Stripping and Grubbing

This work includes the removal of trees, bushes, and topsoil, including roots and other obvious organic matter such as peats. While such work is an obvious first step for any earthwork project, the extent necessary, as determined by reconnaissance and borings or test pits, can be a significant cost factor for the contractor.

Excavating

While excavating is obviously a part of the stripping and grubbing operation, we are here referring to those portions extending to the lines and grades shown on the plans, so the material excavated beyond the organics may or may not be suitable for filling, depending upon its texture. Also, the condition and texture of the excavated surface beyond and below the lines and grades may or may not be suitable. Even if borings have been done, surprises often occur when digging commences. Sometimes the excavating itself can disturb the natural condition of the soil. Thus, it is a good idea to include a statement in the specifications designating a qualified person to inspect and approve the excavated surfaces and the excavated materials to make decisions regarding acceptable condition and texture. Unsuitable excavated material would be designated as waste. Alternatively, depending upon circumstances, it may be feasible and sensible to use some marginal materials as fill in some areas (e.g., lightly loaded, parking areas). Excavated soil of stipulated high textural quality would be used as a structural fill or stockpiled for such use.

Acceptable Fill Texture

What constitutes acceptable texture of a fill depends upon the intended use of the fill, as has been described rather extensively in Chapter 3 and, especially, Chapter 6. Whether it is a select fill to support a pipe, or a clay for the core wall of an earth dam, the specific textural limitations should be stated in the specifications. The usual minimum specification should refer to grain-size limitations for essentially granular soils, typically with respect to limiting the quantities of the very smallest sizes (fines) and the very largest sizes (boulders). For clays, the common basis for defining textural limitations is the plasticity index, but mineralogy and other indices are also used, especially on major jobs such as large earth dams. In such cases, as I have said before and will say again, a geotechnical specialist should be called in.

Representative Samples

A qualified person should be designated to select representative samples from the site or borrow area for purposes of laboratory testing. For textural analysis and compaction testing, this would require about 125 lb of soil taken from the "heart" of the site or borrow mound. If at all possible, a bulldozer should be used to expose a fresh, undisturbed wall of soil from the spot designated by the qualified person. The sample can then be scraped into a bag from a 4- or 5-ft vertical section of the exposed wall. Unless a procedure similar to the one described is used, there is a good chance that the sample will merely be scraped off the ground. Or worse, it will be collected from the base of a slope. In either case, the sample is not likely to be anywhere near representative of the "heart" that will be used in the filling operation, because of the sorting action of rainwater runoff and mass wasting (the accumulation of soil at the base of a slope or cliff, called talus in geologic parlance).

A word about the not-uncommon problem of very heterogeneous soil conditions at the site or borrow area. In such a case, obtaining or even defining a representative sample is an excercise in futility. Until a specific, rational alternate method of establishing field density requirements is developed (such as the Compaction Data Book described in Section 4.3.1), probably the best approach is to specify that an *experienced* soils engineer or technologist be present to inspect and make judgments on field densities. Since the nature of such a job inherently requires many on-the-spot judgments, it is imperative that the person in the field be exceptionally well qualified. The owner should understand, moreover, that certification of the fill cannot be based so much on numerical results but rather on the judgment of the inspector.

Fill Placement

The loose lift thickness should be specified, and it should be stated that the lifts should be spread horizontally. However, it is often helpful to incorporate some flexibility by allowing changes in lift thicknesses as long as it can be demonstrated in the field that the required target value densities can be achieved.

Lift Preparation

Adjustments in moisture content of the loose fill may be required, by aeration (scarification) if too wet, or by spraying with water if too dry. Here, too, rigidity in the specifications should be avoided. For example, the phrase, "the soil shall be compacted at its optimum moisture content" (one which I have seen more than once) is both erroneous and—if taken literally—unrealistically rigid. First, a soil does not have *an* optimum moisture content, as implied by the statement, since the optimum moisture content depends upon the energy of compaction. Second, the chance that the natural soil moisture will be *exactly* at optimum for the energy of compaction being used is remote. Thus, if the specification were to be taken literally, the contractor would have to change either the soil moisture or the compaction energy for almost every lift of soil, an obvious impracticality. I recommend that a range of compaction moisture content be specified. In some cases, notably for clays, a limiting lower moisture content might be stipulated to prevent future expansive behavior. Otherwise, let the target value density control. Remember: end result and suggested method.

On some jobs, mixing two soils before compaction may be in order. In this case, and with soil moisture for that matter, avoid the word thoroughly. Be more specific, or at least substitute "to the satisfaction of the soil engineer" (or other qualified person).

Compaction

While providing suggestions, the stipulation of the exact type of compacting equipment should be avoided. In addition, a range for the number of passes can be suggested, but it is better to stipulate that the methods should be established in the field with regard to equipment, lift thickness, moisture content, and number of passes, specifically in the form of periodic test strips. This is a procedure whereby the inspector and the contractor agree to determine, by a simple trial-and-error procedure, a reasonable combination of methods which will yield the specified end results. Thus, for example, a 150-ft strip of soil can be placed to, say, 9-in. loose lift thickness, and three, five, and seven passes made over three 50-ft lengths. Density tests are made in each section to determine the effects. Thus, early in the job, an agreement is struck on methods to everyone's satisfaction. The use of such a test strip is obligatory in Australia, Finland, and France (Reichert, 1980, p. 188). Incidentally, such a strip can sensibly be within the lines and grades of the plan, to avoid unnecessary extra work. Of course, the methods thus developed will be applicable only if all conditions of the test strip do not change. If the texture of the borrow changes, for example, another test strip and another target value density will be needed.

End-Result Stipulation

The specifications should call for a very specific end result, usually some percentage of the density established by a recognized laboratory compaction test. A method for doing so is described in Section 4.2.3. Readers should note the limitations of the method, particularly with respect to the soil types to which it applies (generally, TALBs). The extent to which the method has been used in practice is not known, so its performance rating or validity is likewise unknown. However, the chart upon which the method is based, Figure 4.2, is one that has been used for many years with confidence for the purpose of assigning allowable bearing capacities for footings on sand.

A suggested sensible approach for selecting percentage compaction is to increase the value obtained by some arbitrary amount, particularly for fills that are significantly different in texture from sand. As stated in Chapter 4, the method should not be used at all for clays. For those who use the method with any regularity, field performance records can be accumulated in order to refine and/or gain confidence in its application. The installation of settlement plates, where feasible economically (i.e., if you can get the client to pay for them), would yield the best data on field peformance.

Rather than stipulating a single end-result percentage compaction, thereby requiring *all* field tests to meet the stated density, it is often sensible to write the specifications "in terms of an average value [of all tests] and provide two levels, one for its acceptance, guaranteeing the quality of the work, the other for its rejection" (Reichert, 1980, Recommendation 7.3, p. 200). Stating such average values and ranges would seem to be most sensible on very large jobs where wider ranges of borrow texture are likely to be encountered.

Consideration should also be given to stipulating how many field density checks should be made (one test per 3000 yards, for example), which field density technique should be used, and at which locations the tests should be done. Specifics pertaining to these features are treated more fully in Chapter 8.

Weather Restrictions

On a fill job of significant size and duration, it is almost inevitable that problems associataed with weather will arise. The borrow may be too wet or too dry because of recent rainfall or lack thereof. Cold weather may create freezing problems, particularly for clays or clayey soils. Precipitation during placement may quickly render the soil too wet for compaction. Finally, unless suitable precautions are taken, the compacted fill can be damaged by moisture changes *after* compaction, but often before the full completion of construction.

Wet Soil.   If the soil is too wet for compaction, as compared to the optimum moisture content for the energy of the compactor being used, the options are (1) to continue operating at less than optimum efficiency, (2) reduce the weight of the roller and use more passes, (3) dry the soil by scarification to an acceptable lower moisture content, or (4) quit for the day or whatever time period will be required for a natural drying of the soils to operable levels.

Which of these options should be followed in a given case and how the decision will be made and by whom are not simple decisions for several reasons. First, there is no exact point at which a soil can be designated "too wet," although obvious undulation of the soil under the action of the moving rollcr is a clear indication that one is beyond the point of productivity. Second, the decision is clearly an important economic one to the contractor, inasmuch as all but option (1) will definitely cost money or time, or both.

However, option (1) may be the most costly if indeed one is just shoving the wet soil back and forth without increasing its density. Third, who decides? Finally, how does one write the specifications to address the problem?

Since my general recommendation is the use of suggested and end-result specifications, the problem of wet soil can be accommodated by language similar to the following:

The soils engineer [or other designated inspector, observer] shall advise the contractor regarding suggested options for dealing with the problem of compacting soil which appears to be too wet for efficient compaction.

In this way, the contractor can decide what to do, fully realizing that the end result (field density, usually) will ultimately govern acceptability.

Dry Soil.   Here again, the designated person can at some point advise the contractor to consider the option of wetting and mixing the soil prior to compaction. Where clay fills are concerned, however, the end result might include the stipulation of a lower limit on moisture content *and* compacted density, to ensure a completed earth structure that is neither too compressible nor detrimentally expansive.

Cold Weather.   Specifications should prohibit the placement and compaction of soils that contain ice and frozen clumps of soil. This would normally be a disqualifying problem when dealing with soils containing silts and clays of significant quantity. Clean, granular soils such as select fills and sand–gravel mixtures may be placed and compacted in such weather, however.

Precipitation during Placement.   Steady rain at the site, if not too intense, can often be handled by specifying (or "suggesting") that "in the event of rain, the soil shall be compacted immediately upon spreading."

Postcompaction Precautions.   The specifications should require the contractor to provide slopes necessary for surface drainage. It may also be advisable to include requirements that will protect against surface damage, notably by freezing, drying, or construction traffic. (In one interesting case, construction workers had warmed themselves with a fire in a 55-gal drum. As still another example of Murphy's law, the spot where they placed the barrel was precisely where a footing was later placed; the baked clayey fill, when subsequently wetted, expanded and cracked the footing, column, and wall above.)

Control

In addition to the usual end-result stipulation of a target value density, other features of fill control may sometimes be included in the specifications. These could include special criteria, matters of fill control procedures, and how various persons are (or should be) involved with implementation and enforcement.

Correction

The specifications should contain language pertaining to the field testing of the fill and correction measures required where tests indicate inadequate compliance. An example of the latter might be that "any lift that is not compacted to the specified density shall be recompacted until the required density is attained."

It may be advisable to include a statement that will have the effect of prohibiting the unsupervised placement of large thicknesses of fill. Obviously, such a feature will only make sense where the contract calls for the continuous presence of an inspector, observer, or other designated person. A clause requiring the contractor to excavate to a level specified by the on-site person should also be considered. This is to guard against "lunch hour filling," a procedure where large fill thicknesses are placed, usually in small, confined areas, and the surface compacted. In such cases, the upper foot or so may meet the target value density, but the lower regions most likely would not. (More about these kinds of sticky problems in Chapter 8.)

Special Criteria. When large or important earth structures are to be built, criteria other than specified densities are sometimes used. Usually, such criteria create more expense and are more technically complex, requiring the attention and involvement of geotechnical specialists. Following is a summary of recent practice, most of which is described more fully in the 1980 International Conference on Compaction (Paris, France, three volumes).

Reichert's review of reports from 32 countries (of 67 polled) reveals (Reichert, 1980, V3, p. 181) that:

The four most frequently used criteria are: the imposition of a Proctor density, the limitation of air percentage, the imposition of a CBR value or of a value for the bearing capacity as given by the plate bearing test.

Of these, the Proctor density is most often used (thus confirming the focus of this book).

Japan and New Zealand impose upper limits of air percentage in the compacted soil, and these limits are typically and logically applicable to essentially cohesive soils. France also uses this criterion for lime-stabilized soils. In Chapter 3, I suggested that the best definition of compaction was the expulsion of air from the soil, so specifying a limiting air percentage is really an indirect way of stipulating a degree of compaction. If too much air remains in the soil, postconstruction loadings will result in rather rapid settlements until 100% saturation (zero air voids) is achieved.

CBR values are used as criteria by Swiss and Thai authorities, generally for road bases and subbases. The CBR (California Bearing Ratio) is rather widely used in the design of pavement support systems and so it would seem quite logical to use the test for compaction control when constructing roadway systems. The test involves the laboratory compaction of a representative sample of the soil in a cylindrical mold, a 4-day soaking period under a surcharge weight simulating field pressure conditons, and penetration with a penetrometer of approximately 2-in. diameter. The CBR is the ratio of the measured resistance (usually at 0.1-in. penetration) to that of a standard crushed rock. The test may be conducted in a similar manner on a prepared secton in the field.

Reichert makes the pertinent observation, as part of his conclusions (p. 198), that

the CBR test should be excluded from certain requirements because of the low diameter of the plunger, the result of the test being very susceptible to the least lack of uniformity in the surface (e.g., presence of a small stone).

(One of my favorite stories along these lines is the consolidation test that wouldn't consolidate. After several puzzling days, the 3/4-in.-thick sample in the ring was extracted, and it was discovered that a 3/4-in.-thick *whole clam* had been neatly encased! Truly a marine clay, if there ever was one.)

Plate bearing tests are used by nine countries. Most commonly, the test is performed for the purpose of determining the subgrade modulus, a measure of the supporting capacity of the subgrade, which is needed to select the thickness of rigid pavements. The test involves the field loading of a 30-in.-diameter plate, with the usual specification that the test is to be made with the soil at or close to its optimum moisture content (Hough, 1969, p. 480). Thus, it is seen that the plate bearing test is not a substitute for compaction testing, but rather an additional control criterion, with, of course, the added cost of the field loading test.

In contrast to the apparent high cost of conventional plate load testing of cohesive soils, a very interesting and promising method of using plate loading tests for controlling compaction of granular soils, rapidly and at low cost, is described by Giddings (1980, p. 547). The method was developed for a large project in South Africa, involving 5 million cubic meters of variable granular material. A conventional tractor was rigged with a front-mounted hydraulic jack and further modified (ballasted) to enable the application of a 2.2-ton load to a test area through a circular steel plate of 30-cm diameter. A guage was used to record settlement to enable the computation of the elastic modulus. Each test required only about five minutes, and the results were immediately available to accept or reject the compaction. Perhaps most interesting is the fact that no laboratory compaction is required. As Giddings asserts:

In the compaction of earthworks, the essential properties of a soil are those of strength and deformation. It is far more logical to obtain a strength or deformation parameter which has real meaning to the engineer, rather than to measure percentage compaction. Percentage compaction is, in fact, only an approximate indicator of the value of these essential properties, established during many decades of experience in varying conditions.

This is the only instance of which I am aware where measuring engineering properties is apparently cheaper than obtaining index properties, in contradiction to Figure 1.2, an interesting anomaly. Giddings is investigating extending and modifying the method, presently limited to granular materials, to include application to more clayey soils. It would seem that this would be a very worthwhile endeavor, especially for possible applications to very large earthwork jobs where field research and the development of specialized equipment can be justified economically.

However, a question arises regarding whether or not Gidding's plate loading control method is generally applicable, even to granular soils, specifically because of the static nature of the test. It is well known that rather loose granular soils can sustain substantial static loadings but will settle dramatically under the influence of vibrations. In Section 6.5.2, I described and suggested a simple experiment using a 1000-cc graduated cylinder. After placing sand loosely in the cylinder, I suggested that a static load—even a substantial one—would not induce a significant settlement. Having performed this experiment many times as a classroom demonstration, I can report that the load was my own weight; I rather awkwardly chinned myself on a piston. I then had one of the students confirm that little settlement took place under the influence of the static loading. A hammer was then used to demonstrate the dramatic effects of vibration on the settlement of the sand (about 10%). I have computed the static pressure in the cylinder experiment and, by a remarkable coincidence, the pressure is almost exactly the same as that used in Gidding's plate loading test, 40 psi (200 lb/5 in.^{2}; 2.2 tons/109 in.^{2}).

Figure 7.1 illustrates what appears to be a contradiction to the assertion inferred by my cylinder experiment—that the settlement (or elastic modulus) should not be significantly affected by the density of the soil in a static test. I believe the apparent contradiction can be explained on the basis of differences in soil mineralogy. The soil in the South African project is described by Giddings as of "a porous nature. . .of weakly cemented material. . .[which would be] "broken down by laboratory compaction." The field compaction was by "very heavy vibratory compaction machinery." For the plate loading tests of Figure 7.1, compaction was done by "a towed single drum vibrating roller of 14 tons static mass" (Giddings, 1980).

What I believe happened was that the static loading of the plate bearing test induced structural breakdowns of the weakly cemented soil particles at the intergranular contact points within the zone of influence (pressure bulb) of the loaded plate, most of which occurred when the soil was in the relatively looser conditions (lower number of passes). In the case of the cylinder experiment, admittedly a very crude test, the loose high-quality sand was able to resist the static pressure without interparticle breakdown,

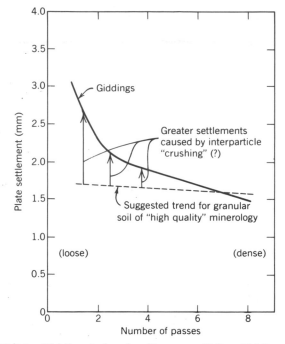

FIGURE 7.1   Giddings' plate loading tests. (After Giddings, 1980.)

and no significant settlement occurred. (The sand used in the cylinder was Ottawa Sand, the same type used in the conventional sand-cone field density test.)

If my speculation is correct, a similar field experiment on a high-quality granular material (of sand sizes) would result in a much shallower curve, irrespective of density, and that even a loose soil would give the illusion of denseness because of the static nature of the plate loading test. Fortunately, the construction method involved six passes of a heavy vibratory roller, so the quality of the completed earth structures was assured. My question concerns whether or not the method of control applies in general to the compaction of granular soils, for there would be a danger of accepting loose soils based upon static testing, only to have postconstruction vibrations induce intolerable settlements. There is a classic, famous case history that illustrates this concern. Terzaghi was called in to investigate the settlement of a series of structures that had stood for many years without movement or distress. With his unparalleled skills of observation and of assessing cause and effect, he determined that loose sands underlying the row of structures had started to settle under the new influence of heavy truck traffic on the road upon which the structures stood. The clue was that all of the structures were tilted toward the road. Apparently the vibrations were sufficiently damped so that there was little effect on the sand toward the rear of the structures.

Hendron and Holish (1980, p. 565) report a method of compaction control of a major earth dike system of 3 million cubic yards using cohesive soils with widely variable properties. Their approach was to augment density testing (laboratory and field) with the development of correlations between direct measurements of degree of compaction and various index properties of the soil, including liquid limit, the fines content, and the unconfined compression strength as measured with a pocket penetrometer. This major field effort was interestingly similar to the approach I describe and suggest for the problem of "Changing Borrow" (Section 4.3).

Eggestad (1980, p. 531) describes a method of compaction control that he developed and field tested called a comprimeter, a penetration apparatus based upon the common-sense principle that a rod driven into a dense soil will cause heave of the surrounding surface, and that the heave volume would be directly related to the degree of denseness (including volume reduction for very loose soils). The device was compared to the more common control methods of field-density testing: the water-balloon and nuclear device methods. The results showed that the most rapid and simple method—the comprimeter method—gave very reasonable results for sands and sand–gravel mixtures. Each field measurement took less than 5 minutes. In his conclusions, Eggestad makes the salient observation that "compaction, even in small earthworks should be controlled."

Finally, Reichert reports that other control methods are sometimes used when conventional tests are not feasible; for example, proof rolling of rock fills where the sizes of the rock preclude tests such as CBR or plate bearing.

Fill Control Procedures.   Depending upon circumstances, and perhaps especially on how far along the project has been developed in terms of planning, it may be advisable to include some language in the specifications covering fill control procedures. As stated under the end-result stipulation, the number of field density tests, and methods for controlling (choosing) their locations should be stated. But also included might be such things as methods of testing: field-density testing (water-balloon, sand cone, or nuclear), tests for texture (sieving, plasticity), or the specifics regarding special field testing criteria such as CBR, plate loading, or full-scale tests for special circumstances.

Because one of the purposes of this book is to provide what amounts to a manual for such work, the full details of these procedures is consolidated in the following chapter.

7.2   Implementation and Enforcement

Here we come to a very difficult and sensitive aspect of fill control or, indeed, any type of engineered construction: How do we go about implementing and enforcing the features of the contract documents, including the plans and specifications? Corollary questions would include: Who has the responsibility(ies)? The authority? Who gets the blame when something goes wrong? Easy questions, tough answers!

Aggravating the problems considerably are some factors that have evolved and become disturbingly pronounced over the past 10 or 15 years. One is a broad, national trend in American life. The others are direct results, I believe, of the national trend.

We have become a litigious society. The number of American lawyers increased by 83% in the 1970s. By contrast, Japan produces twice as many engineers as we do with half our population base. In fact, there has been a decline of between 5 and 10% in American scientists and engineers during the same time period. Now the American Bar Association and the American Civil Liberties Union may not perceive these data in the same light as those involved in engineered construction, but no less an authority than the Chief Justice of the United States Supreme Court has spoken out recently against the proliferation of lawsuits and the resultant backlog in the courts. Certainly there are many good and proper instances where a person needs legal recourse to resolve a problem, but it would appear that a serious imbalance has developed.

One incident from my own fairly recent experience comes to mind. I had been retained by a law firm as a potential expert witness dealing with a foundation failure. Before the case got to trial, during the deposition phase, I met with a lawyer from the firm that retained me. He confided to me a modus operandi whereby one cites as many defendants as one can possibly identify, however remotely involved with the construction failure, so that there will be more potential contributors to the settlement pot. (For those who are not aware, negotiations for a monetary settlement go on continuously before and even during the trial. On one case, the settlement occurred in midtrial. I suppose it depends during trial on the perceptions of each side of how the trial is going, and which experts have been most effective with the jury.) Upon later reflection on this revelation, I realized that I could be one of the defendants cited in some future lawsuit, particularly if the modus operandi is indeed common practice. The insidious closure to this loop is that you would have to hire a defense attorney, and the expense may be more than it would cost to contribute to the settlement. Thus, a completely innocent person can be damaged financially and his reputation tarnished by the stigma of settlement.

A further effect of this litigiousness is reflected in the decision of an academic colleague of mine to decline part-time design consulting work because of a legitimate fear of being dragged into a lawsuit in the manner described. The cost of professional liability insurance has gone up in accordance with the litigation explosion, and the small volume of work that a typical engineering professor has time for does not generate sufficient income to warrant the expense of the insurance. The man in question is a highly competent, well-educated, licensed, professional engineer (M.S., structures; Ph.D., geotechnical; postdoctoral, Fulbright). Thus, the public is partially deprived of the contributions of one of its better engineers. For many years I did small consulting jobs without liability insurance, figuring my competence would serve me well enough as protection. I now see that that is not necessarily the case. I know of other part-time consultants who operate without insurance, but as they say in Atlantic City: "It's a crap shoot." Of course, full-time practitioners operate with full insurance protection for themselves and their employees. If they don't, they'd better! These costs are, of course, passed on to their clients. Joe Public always pays the bills!

If any further proof of this national trend is needed, consider this assertion: "As a field engineer, you *will* appear before a court of justice during your career!"

This quotation (emphasis mine) is from the first chapter, first page, of a book entitled, *Sue the Bastards: Handbook for the Field Engineer,* by Frederick Richards, P.E., 1976. The paperback was sponsored by the Rochester Section of the American Society of Civil Engineers. The startling cover, which goes along with the graphic title, is a photograph of a hard-hat worker in a jail cell! Interesting if a little scary reading.

The effect of litigation fear (as Woody Allen says, "you're not paranoid if people are *really* out to get you.") in field work is the introduction of a new breed of field worker: the *observer*. I do not know when this person came into being, for there were none in the early 1960s when I was engaged in fill control work as an *inspector*. I suspect that some bright lawyer made a convincing semantic link between inspection, responsibility, and culpability, won the case, and thus established a legal precedent along those lines. The response was the birth of Omar, the Observer. I have chosen the name of this fictitious person for two reasons. First, it is alliterative, and I've always liked that. Second, I don't know anyone by that name, and I'd hate to lose a friend or even offend a casual acquaintance. For Omar is somewhat scatterbrained. He is the fellow about whom everyone asks, "Who hired him, and why was he hired?" (More often than not, he was somebody's nephew.)

Mostly, I heard about Omar through comments by graduate students who were working full time as engineers in training. Lacking first-hand experience with this development, I have talked with colleagues in professional practice who have dealt with this new feature of construction practice at the management level and have solicited their opinions of its ramifications. That is the basis of what I report to you here.

What I suspected, and has been confirmed, is that observers serve only to exacerbate the problem, adding more confusion and more cost to an already confused and costly situation, principally by making it more difficult than ever to establish clear areas of responsibility and authority. As T.B.* put it, "anyone who accepts responsibility without authority is crazy."

To the extent that the term inspector may suggest authority, the des-

*See acknowledgments.

ignation observer seems to connote impotence. In a hypothetical but potentially very real scenario, an observer might report, "I told him the fill was no good, but he poured the footing anyway."

T.B. would respond, "What do I need you out there for; I could put in a television camera."

What it amounts to is that, somewhere along the line, the buck has to stop, sensibly with someone in the controlling agency. For example, a professional engineer will have to state in writing that "the specifications for the fill have been complied with" and affix a seal accordingly. It would seem that the use of the term observer in engineered construction of any sort is a disservice to all involved. First, as I have noted already, it is common that young, relatively inexperienced people are currently assigned to fill control field work (although this may change with the advent of construction engineering technology curricula). To further saddle them with the designation observer is, in my view, ill-advised, for it will further reduce their potential for effectiveness from a tenuous position to one of virtual powerlessness. The client, of course, will not be too happy with prospects for getting value for his money. The engineer, or whoever must certify the work in writing, will not be able to do so confidently if the field person has not been able to exercise any authority on a day-to-day basis. And finally, the contractor will eventually suffer the inevitable consequences of performing without the daily pressures of reasonable and effective control procedures. (While some contractors may scoff at this, I am convinced that the best chances for success and profit spring from long-range reputation for the quality of completed structures that are free of postconstruction problems. I further believe that it is the nature of us all that we need a *reasonable* degree of steady pressure to operate effectively—that includes inspectors, engineers, and contractors.)

I would thus discourage the use of the term observers in contracts and specifications. Instead, I would encourage the return to the use of terms such as inspector or controlling agent, and strengthen the line of control and authority by inserting appropriate language in the contract documents, language that would carefully define responsibility and authority. For fill control work, I suggest the following as a guideline for terminology and tone:

The inspector shall advise the contractor of the rejection of any unacceptable end results, reiterate steps toward correction as may be stipulated in the specifications, and (optionally) advise the contractor on alternative methods of obtaining the end results required.

7.3 *Nontechnical Aspects of Specifications*

Fletcher and Smoots (1974, p. 390) stated that specifications should set forth all the requirements of the work to be performed. I believe that I have included all the technical aspects of compaction specifications in the preceding sections of this chapter, and that is the stated focus of this book. There are, however, other aspects of specifications, not yet mentioned, which should be included. These include (Fletcher and Smoots) "terms of payment, time allowed for completion, and provisions for adjustment for changes in the work.") Such feaatures are clearly of vital importance to the client and the contractor.

7.4 Specification and Project Evaluation

Since this book is written largely for nonspecialists, but including the young field engineer and engineering technologists (collectively, the inspectors), most readers would be in the position usually of being given a set of specifications for whatever their various purposes might be: the client perhaps to see what he is paying for, the contractor to bid the job, the architect or structural engineer because of their indirect but important involvement with earthwork operations, the construction engineer because of the need to advise the contractor directly, and the inspector because of very direct involvement with control. Thus, the typical need of such persons would be to evaluate specifications, rather than write them, and to otherwise review all aspects of the project prior to the start of construction.

One of the more obvious ways to do this is simply to read the preceding sections of this chapter and make an item-by-item comparison of your specs with what is described here, in a checklist fashion. I recommend this procedure, but would also offer some additional guidelines, for sometimes one can get so bogged down in details that one fails to see the forest for the trees. Group your evaluation process into two categories: soil and equipment.

7.4.1 Soil: Quantity, Texture, and Density Requirements

1. Is the quantity of soil (or rock) sufficient for the volume of fill needed for the site preparation work?
2. Is the texture of the soil (or rock) that is available in accordance with the specifications?
3. What will be the acquisition problems and costs?

These questions must be considered together. If the site preparation involves cuts, as it usually will, will the excavated soil be of sufficient amount and of suitable texture? You'll need to have borings, test pits, or (at the very least) some soil maps describing (probable) soil texture to get the answers needed. Soil maps for most counties of the United States are available through regional offices of the Soil Conservation Service (SCS), a division of the U. S. Department of Agriculture. Maps and accompanying text describe soils' texture to a depth of about 10 ft and also provide information on depth to bedrock, drainage characteristics, and depth to water table.

For large jobs, airphotos may be very helpful in locating potential borrow sources, especially where soil maps are not available (e.g., outside the United States). Air photo interpretation is the art of identifying geologic landforms through the study of clues such as drainage patterns, gully erosion, tone, vegetation, and land use. Once the landform has been identified, the texture of the soil in the deposit can be predicted within reasonably well-defined limits. A limited number of borings or test pits may be planned for purposes of confirmation. However, other acquisition problems should be considered simultaneously or in some logical, sequential fashion. Who owns the land? Will they sell it? How much? What is the haul distance? Will there be access (trafficking) problems? How much stripping and grubbing will be required?

4. Are the density requirements sensible and specific? If the job is large and complex, there may be several different types of soil texture needed, each compacted to densities compatible with their intended use. The end-result densities should be in terms of specific laboratory test names. (Standard Proctor, modified Proctor, relative density, and/or their correct ASTM designations.)

In going through this evaluation process, the nonspecialist should get a feel for whether the specifications are well written and complete. He or she should get a sense of whether the specifications are what Fletcher and Smoots call "a cut and paste" document from other jobs. Finally, one will more than likely be able to decide whether a specialist's services are needed. If so, this is my best advice: Call one in before it's too late.

7.4.2 Equipment: Excavation, Loading, Hauling, Placement, Precompaction Preparation, Compaction

Is all the equipment available to do the job, and is there a reasonably good balance of the different kinds of equipment? The logic of the title of this section, and the broad question stated above, is based on a consideration of the chronological sequence of operations in any earthwork construction: the soil (rock) has to be excavated, loaded, hauled, placed, prepared (sometimes), and, finally, compacted. As a simple example of the balance referred to in the question, it can sometimes develop that one has too much placement equipment relative to one's compaction capability. (In a case history that I describe in Chapter 8, this happened to an almost laughable degree.)

Excavation can, of course, be accomplished with an extremely wide assortment of equipment, from hand shovels to blasting. Not only is the amount of excavation of concern, but the ease with which it can be done is of major importance in terms of equipment selection and cost. Indeed, I would venture an educated guess that the accuracy of estimating the quantity and ease of rock excavation can, and often does, make the difference between profit and loss on a project. As one contractor stated: "When the estimate of rock is out 50%, that's not too bad. When it is out 100% we can live with it. But when it is out 10,000%—as on our last job— somebody gets hurt" (Greer and Moorhouse, 1968). In addition to determining the amount of rock excavation, the question of whether the rock will require blasting or ripping is a major cost factor, and, or course, the selection of equipment cannot be made without first making this determination. A common means of doing this is the use of a rippability chart based on seismic refraction exploration. (Schroeder, 1980, p. 109–110)

To a lesser degree, the ease of excavation of soils can be important. For example, if all excavation is to be done on or very close to the site, and the soil is loose or soft, one special-purpose machine—the self-propelled scraper—may be used for three operations: excavation, hauling, and placement. (A scraper is a machine designed for the sole purpose of picking up soil from one place and spreading it in a lift where required. It is not an over-the-road vehicle.) If the soil must be hauled from a borrow area remote from the site, involving transportation over public roads, then a wider assortment of equipment will be needed. Even when only on-site excavation is required, however, the scraper may not have the power necessary to excavate soil that is dense and of significant cohesiveness. Glacial tills, for example, typical in northern sections of the United States, are almost always difficult to excavate, requiring power shovels or similar expensive equipment. If self-propelled scrapers are used, a bulldozer will commonly be required to assist the scraper by pushing. Alternatively, a tractor-drawn scraper may be used. Overlooking this need, particularly on a large earthwork job, could result in significant loss (or reduced profit) to the contractor.

The best type of equipment for placing soils is the scraper. An experienced operator can eject the soil from the "bowl" of the scraper in a controlled fashion to spread the soil in a thin layer over a strip, typically in accordance with the lift thickness stipulated in the specifications or as modified by performing test strips in the field.

Soils may also be placed by some type of over-the-road haul unit (i.e., a dump truck). This would normally be the case when the soil must be trucked in from an off-site borrow area. In such a case, the dumped soil must be spread, usually by a bulldozer, to the required lift thickness before compaction (or any required precompaction preparation).

Tank trucks with sprinkler bars, or some type of equipment for drying by scarification, may be required to adjust the moisture content of the soil.

The final step is, of course, compaction. Guidelines for more detailed evaluation and selection of equipment for excavation, hauling, and placement of soils are provided in a recent, thorough treatment of the subject, *General Excavation Methods* (1980), by A. Brinton Carson, General Contractor and Professional Engineer. To augment recommendations and other sources given in this book, I also recommend Carson's lists and discussions of different types of compaction rollers (Chapter 11).

7.4.3 *Compliance with Applicable Laws and Regulations*

In evaluating (or writing) specifications, one must be certain that there are no conflicts or omissions with respect to applicable building codes, ordinances, or an assortment of other regulations, increasingly of an environmental or safety nature (Occupational Safety and Health Administration, OSHA).

7.5 Glossary

Settlement plates.   Field instrumentation of a completed structure that permits periodic readings, using an engineer's level, for the determination of settlement and settlement rates. Although the term plates is used, the instrumentation can take any form, as simple as a scribed horizontal line on a wall. For fills, a horizontal plate, with an attached riser pipe extending out of the fill, is commonly used.

Elastic modulus.   The slope of a stress–strain curve. If the material is elastic (as steel), this slope is a straight line. If the material is other than elastic, as with soft soils, an appropriate representation of the stress–strain curve is determined; usually then called a *subgrade* modulus (as in field plate bearing tests).

Proof rolling.   Rolling a soil or rock surface with a very heavily loaded compactor for the purpose of locating loose or soft areas; "troubleshooting."

Geologic landforms.   A distinctive form, often in both shape and texture, associated with a particular geologic process. Once the landform is identified (e.g., by aerial—or even satellite—photography), much useful information about texture and potential foundation suitability can be predicted. In geology, the study of landforms is called *geomorphology*.

8

Fill Control Procedures—
Inspection

After all is said and done, the bottom line (as our lawyer friends say) is "will the job get done as planned so that no post-construction distress will occur?" Assuming that the specifications have been well written, this will depend to a large degree on whether or not the specifications are followed by the contractor. On typical large jobs, where extensive filling is required, a necessary ingredient in compliance is enforcement by the controlling agency, followed by written certification to the owner. Ideally the controlling agency would be one with expertise in geotechnical matters such as a consulting firm or a highway department. The direct controlling agent is, of course, the field inspector. This chapter is largely directed to the latter. To a lesser extent, it is also directed to contractors in recognition of the fact that there will be many smaller jobs, or special circumstances, where no inspector will be present, so that compliance with specifications (or good construction practice when there are no specifications) will be based on pride in workmanship and reputation for excellence—a priceless commodity for a contractor (or anyone else, for that matter!).

There is even something here for structures specialists.

8.1   Field-Density Testing

Assuming homogeneous soil conditions, if all footings of a structure have the same bearing pressure, will there be any differential settlements between footings?

Think a bit. *Before* checking the answer! The answer is at the end of the chapter.

8.1.1   The Pressure Bulb

I started this chapter with a quiz to illustrate a common serious misconception, particularly among structures types. Although I have not compiled records, I can say with certainty that a large majority of engineers and engineering students will answer the question incorrectly. I think I know why, and it is partly the fault of those of us who write textbooks. The source of the problem is that bearing capacities are specified in tons per *square* foot (or similar units). The use of planar units leads a reader to think only in terms of the pressure that exists between the footing and the soil—the contact pressure. What governs the settlement, however, is, of course, the *volume* of soil that is stressed significantly, hence the fundamentally important concept of *the pressure bulb.* There are, in fact, an infinite number of pressure bulbs, as revealed by theoretical studies of stress distributions in semiinfinite masses. The ever-pragmatic engineer, however, asks the mathematician, "To what regions do significant stresses extend?" To which the mathematician responds, "What's significant?" Engineer: "Oh, say 10% of the contact stress."

The answer is shown in Figure 8.1, which depicts a loaded area of width *B* exerting a contact stress *p* on the homogeneous soil. The $0.1p$ pressure bulb may be thought of as the standard pressure bulb, or better yet, the rule-of-thumb pressure bulb. There are a number of reasons for the latter nomenclature. First, the shape of the loaded area in application would vary, typically square, rectangular, or long for footings, and elliptical for wheel loads. The depths of the corresponding $0.1p$ pressure bulbs for each shape would be different. Hence, most engineers think roughly "twice *B*" as the rule of thumb for evaluating the general effects of stress distribution. It is this line of thought that probably led to the following recommendation for soil boring depths:

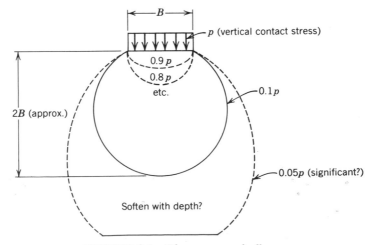

FIGURE 8.1   The pressure bulb.

Beneath lightly loaded structures with widely spaced columns, the depth of the first boring should not be less than twice the probable width of the largest footing.

(Peck et al., 1953, p. 139). Other reasons for using the figure as a guideline rather than a rigid rule are (1) p itself may be unusually large, and hence $0.05p$ may be significant, or (2) the soil below the $0.1p$ pressure bulb may be unusually soft, so that even very small pressures may induce major settlements. Geologically, softening with depth occurs either when the water table has been lowered, or when the entire physiographic region has risen because of tectonic forces. In either case, the region near the surface dries out and forms a crust that is stiffer than the soils below. Such a crust is common in the New Jersey Meadowlands. Conversely, if a number of borings have indicated that bedrock is shallow beneath the proposed structure (i.e., considerably less than 2B) it would make no sense to follow slavishly the rule of extending borings to the 2B depth. "2B or not 2B, that is the question."

Finally, no soil or rock is homogeneous. Indeed, some have very pronounced layering, as with the varved clays of the Meadowlands, and stress distributions are further affected in a complex way.

To further illustrate the importance of using the pressure bulb to guide one's thinking, I offer a story. When I took my orals for my doctorate, one of the professors on my committee asked what I would consider to be the limiting allowable contact pressures on a typical state highway. Since I had taken a course in pavement design with this professor, the question did not surprise me, and I answered "about 100 psi" by interpolating between a typical automobile (30) and airplane tires (200). The follow-up question was significantly more puzzling:

Mr. Monahan, are you familiar with a current fashion among the young ladies called "spiked heels"?

I indicated somewhat hesitantly that I was.

What would you estimate the dimesions of the contact area of a heel?

Answer:

"About ¼-in. square."

Next question:

Do you have any preference for stature in young ladies, Mr. Monahan, petite versus hefty?

Monahan (getting nervous, now):

Petite.

What would the contact pressure be for, say, a 100-lb young lady, Mr. Monahan?

(Wielding the chalk, the quick calculation revealed the answer.)

"Uh . . . 1600 psi."

The final query:

Can you explain, Mr. Monahan, why the state of Oklahoma places a limit of approximately "only" 100 psi on wheel load contact stresses, in view of what you have just calculated?

(Wow! Gulp!) Sure! Fortunately for me, I had had a few years teaching experience by then, which helps one to think better on one's feet, so with a minimum of apparent distress I searched my brain and asserted,

Why certainly, it has to do with the size of the respective pressure bulb and the manner in which the stresses dissipate with depth!

Since that time, a quarter of a century ago, I have related the story in each of my soil mechanics classes, in the belief that in a graphic, humorous, and virtually *unforgettable* way, the concept of pressure bulbs was permanently implanted. One of the funniest exchanges I ever had in class came about when a student offered the "correction" that the contact pressure of the spiked heel should be 800 psi, assuming of course that the girl had two feet, to which I replied, "Only if she hopped like a bunny . . . and in that case, impact load would have to be considered."

The applications of the concept are extremely important and varied. Hopefully, one would now immediately "think pressure bulb" whenever one sees or envisions a loaded area (or areas), rather than make the serious mistake of considering only the contact area stresses (the typical error of structures types specialists). Figure 8.2 shows some which come to mind.

Parts *(a)* and *(b)* represent, to approximate scale, the spiked heel and a tire, respectively, bearing on an 8-in. pavement. It can be seen that the bulb of very high stresses of the heel is only about ½-in. in diameter and that the stresses dissipate "quickly" to nominal levels within the zone of the pavement. By contrast, the stresses associated with the tire extend significantly into the base course and (not shown) into the subgrade (natural soil) below. One can see from these simple considerations that pavement design is an exercise in providing a series of materials (pavement, base courses) of decreasing strengths and stiffnesses, top to bottom, corresponding to the diminishing stresses that will be imposed by the surface loads.

Figure 8.2*c* is a generalized sketch of any standard foundation element, an isolated footing (one of many supporting a building), an entire building (as for a mat foundation), an abutment, or a pier. The *B* and 2*B* dimensions

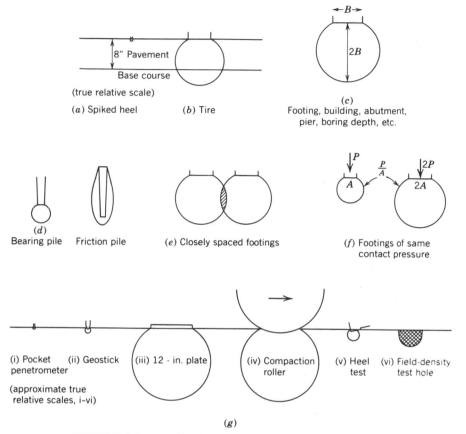

FIGURE 8.2 Applications of the pressure bulb concept.

are shown merely to suggest a visual device for assessing the zone of influence of significant stresses, a very valuable tool for guiding one's judgment in making decisions on the needs for sampling and laboratory testing throughout the duration of any project, from initial conception, to planning the first day's boring program, through the months of modification of one's thinking (a process probably unique to foundation design), to the final selection and detailed design of the specific foundation scheme.

Figure 8.2d simply extends the notion to pile foundations. Part (e) shows two closely spaced identical footings, illustrating overlapping pressure bulbs. Because of the higher stresses in the zone of overlapping, one would anticipate the probability that the footings would settle unevenly, tilting toward one another.

Figure 8.2f provides a definitive answer to the question at the beginning of the chapter. Even though the contact pressures are the same for each footing, consideration of their pressure bulbs leads one to the (now) obvious conclusion that the larger footing will settle more than the smaller, simply because a much larger volume of soil is stressed significantly.

Now we consider specific tools and techniques that relate more directly to fill control operations. These are represented by Figure 8.2g. Part (i) depicts a pocket penetrometer, a device that has a spring-loaded plunger or needle about ¼ in. in diameter. One merely pushes the device into the soil and, in seconds, obtains a reading of the unconfined compression strength, sometimes also called the presumptive bearing capacity for clay soils that are to be loaded rapidly. Notice the specific limitations; it applies only to clays and only to those that are to be loaded rapidly. (Rapidly in this case refers to routine construction, measured in months, where no special surcharging and preconsolidation—usually by the installation of sand drains or wicks—is to be done.) But consider the other unstated limitations related to the tiny size of the pressure bulb ($2B = \frac{1}{2}$ in.). Add to this the potential and probable disturbance of the zone of the pressure bulb by construction machinery or by desiccation, and one wonders why the term presumptive is used. The dictionary (Webster's) states, "giving reasonable ground for belief." I would say, "Don't believe it!"

If the use of the device is extended to fills that are other than 100% clay, the presence of a fine piece of gravel (say 3 mm) or even a coarse particle of sand (1–2 mm) could affect the result considerably. A pebble? Forget it! Part (ii) of Figure 8.2g shows a larger penetration device, the geostick, available from the Acker Drill Co. It is about 3 ft long, 2 in. in diameter, and has a removable (threaded) cone tip. The handle end is a geologist's pick (and hammer). The tube is hollow and can be used to sample clays (although probably not very stiff or hard clays). The penetration feature of the tool enables one to obtain the (presumptive?) bearing capacity, again in seconds, by simply penetrating the soil vertically with the cone attachment. For soft clays, this is accomplished with the weight of the stick; for stiffer clays with the weight of the user. There is a series of circles on the cone representing increments of penetration. Printed on the barrel is a table relating penetration values and bearing capacities for stick weight, 140, 160, 180, and 200 lb. The geostick is a pretty handy device, especially since it has multiple uses. Used properly and with good judgment, it is probably superior to the pocket penetrometer, if for no other reason than the larger size of the pressure bulb which it generates. I would also suspect that it can be more readily extended to fill control work on soils other than clays, though not to soils containing large quantities of coarse to medium gravels. The limitations that apply to the pocket penetrometer also apply to the geostick, but logically to a somewhat lesser degree. Still another type of penetrometer (not shown) is a Torvane device. This is a pocket tool, the main feature of which is a series of blades or vanes. It is inserted by hand into the soil and twisted until the soil fails; the unconfined compression strength is then read directly from a dial. Again the device is largely limited to clay soils, and judgment must be used in extending its use to others. One advantage of the Torvane is that it can be used to obtain readings from vertical or sloped soil surfaces, which would be problematical with the geostick.

As described in Chapter 7, special criteria such as plate bearing tests are sometimes used in fill control. Because of their expense, these are usually limited to very large earthwork projects. Circular plates are used ranging in diameter from 12 in. to 30 in., the latter size being a common standard. In some cases, up to three plates of varying size are used to permit more valid extrapolation of the results to the probable performance of the full-size structure (Johnson and Kavanaugh, 1968, p. 347). For pavement design, the 30-in. plate is usually used to determine a subgrade modulus (Yoder, 1959, p. 337). In addition, special tests are devised for specific purposes and applications, such as for the job in South Africa described briefly in Chapter 7. In all cases, the interpretation of load–settlement curves is necessary to obtain whatever answers are sought. Figure 8.2g(iii) shows the pressure bulb. Applying the 2B criterion, one can see that the pressure bulb can be up to 5 ft deep (for 30-in. plate). Contrast this to the pocket penetrometer! (But also contrast the cost!)

Part (iv) represents a field compaction device and its pressure bulb. No scale is used, for it is intended that the sketch represent anything from an air-operated hand tamper (for ditch work, typically), to a standard roller, to an off-road supercompactor capable of efficient compaction to depths of as much as 10 ft.

Before proceeding to a description of the details of direct field density testing, one last pressure bulb should be mentioned, that of the "tried and true," historic, traditional method of fill control, the famous *heel test* [part (v)]. This test is normally most commonly used in the early phases of a fill control job, in what might be called the getting-things-squared-away phase, before methods have been established that will bring about the desired and specified end-result density. Usually the inspector arrives at the site, and the placement and compaction process is observed. The first thing one does is check the thickness of the loosely placed lift, usually by inserting an opened section of a 6-ft folding rule. Nodding affirmatively (assuming correctness), one then stoops over and obtains a handful of the soil, inspects its texture by rolling the soil through the fingers, looking sort of thoughtful though noncommittal in the process. Then one squeezes the soil firmly in the hand to check for moisture. The compaction rolling is observed, and in some way it is made plain that the number of passes is being recorded, perhaps by holding three fingers up during the third pass. Upon completion of the agreed-upon (or specified) number of passes, it is time for the heel test, which is simply the act of kicking down on the rolled surface several times, accompanied by appropriate facial expressions and head movement (side to side, of course, denoting dissatisfaction). I am being facetious in describing the scenario, but I believe the description is reasonably accurate and that the scene has been enacted many, many times without too much deviation. It's a game which is played over and over, even though the heel test appears nowhere in specifications or (until now) in the literature of earthwork construction, the test does have its serious implications. First (and I am here addressing the young inspector), don't ever use it on the surface of a rolled fill that you did not see placed and rolled, especially if the fill is in a confined area such as backfill for a wall or trench, for surface rolling could compact the soil adequately near the surface but have little or no effect on the soil below. In fact, if the soil is at all clayey, a half hour of baking in the sun could dry out the soil and create a very hard, desiccated crust that would give misleadingly good results in the heel test. I know of a case where a fill was approved on the basis of comparable inspection, a footing was built on the fill, and postconstruction settlements caused serious column and floor damage. Consideration of the pressure bulb associated with the heel test will reveal the dangers of placing one's trust in its validity as a sole means of fill control. The same can be said of more sophisticated tests such as the pocket penetrometer, the geostick, and even 30-in. plate bearing tests of a 10-ft dumped fill. In a word, there is no substitute for observation *augmented* by a carefully conducted field testing program, sometimes utilizing a variety of testing techniques and tools, some for direct measures of density and others (penetrometers) for developing correlations that may be used to complement the control process. As confidence develops in the complementary methods, they may be used as temporary substitutes when time pressures and unforeseen circumstances force periodic suspension of direct measurement.

Part (vi) of Figure 8.2g represents the field-density hole, by far the most widespread technique for checking the adequacy of compacted fills.

8.1.2   Density Tests

Field-density determinations can be made in a number of ways, including the sand cone method, the water balloon method, the jacked sample method, the chunk sample method, and through the use of nuclear devices. All but the latter are destructive tests in that they require obtaining a sample of the soil, weighing it, getting its moisture content, and determining its volume.

The Sand Cone Method

Figure 8.3 is a photograph of a field-density test in progress using the sand cone method. End-result field densities are designated in specifications as some percentage of a laboratory determined maxiumum dry density: the target value. As explained in Chapter 3, the dry density is computed from three variables: the weight of the soil in the hole, $W;$ the volume of the hole, $V;$ and the moisture content of the soil, $w.$

To conduct the field test, one first selects the spot and elevation to be tested (more on this later), and then prepares a fresh surface of the compacted fill for testing. As a minimum, this usually involves scraping off the top few inches of the rolled fill to eliminate the artificial effects of shrinkage by desiccation. (This may or may not be covered by the specifications; perhaps it should be.) Typically this might require tedious hand work with a shovel and other tools. In the photograph, the surface was prepared by passes of a bulldozer blade, at my direction. A level, smooth spot was chosen

FIGURE 8.3   Sand cone field density test.

(obviously between the tracks of the treads), and the plate set down in a firm, flush position. (With extensive amounts of gravel and cobbles present, as was the case here, finding a spot to test is not always as simple as saying it. Even then, the content of the hole may invalidate the test, as we will see.) The density hole was dug in a roughly hemispherical shape, and the contents placed in the can for purposes of determining the soil weight, W, and its moisture content, w.

Before doing so, however, one must make a qualitative judgment on the acceptability of the contents of the hole in terms of texture variation compared to the laboratory test borrow. Remembering that the essence of valid laboratory testing is to *simulate field conditions*, the soil in the hole must reasonably conform in texture with that which was tested in the laboratory. To define reasonably, what I am saying is that you should not have a 3-in. cobble in that can, such as the one shown in the left foreground. If you blindly accept such a "dig," it is predictable that the computation for the density will turn out to be about 160 pcf, the density of solid rock being about 170 pcf. (Moreover, if you do accept the test, and report its results to your project engineer, he will wonder about your wrapping.) In Chapter 3, it was noted that the laboratory test requires the exclusion of larger particles, depending upon the size of the compaction mold used (see Section 3.4, ASTM Compaction Requirements), and that the percentage of the excluded material is recorded. Assuming that the soil from the test hole reasonably conforms to the texture of the borrow, what do we do about the inevitable difference between the percent ($+n$) excluded in the laboratory test compared to the corresponding percent in the test hole? The solution: an empirical correction. (The formula that was used is proprietary.)

Returning to the field density test, it remains to explain the techniques used for determining W, w, and V.

The weight W is determined straightforwardly by weighing the can of soil and subtracting the tare weight of the can. Typically, the field inspector will work off the tailgate of a vehicle and will be supplied with a standard triple beam balance of rugged design and suitable capacity and sensitivity. To the young inspector: Learn how to use it. Don't forget the hanger weights.

The moisture content w may be determined by "cooking" or by the use of a "speedy moisture" device. The cooking technique is the same used in the laboratory, except that one would use a Coleman camp stove or similar instead of the usual electric oven of the laboratory. For convenience and to improve accuracy, the entire can of soil can be cooked to constant weight, and the moisture determined by before/after weighings. The speedy moisture technique involves the use of a commercially available device utilizing a measured powdered chemical that is mixed with the moist soil in a chamber, producing a pressure. A gauge is attached that is calibrated to read moisture content directly.

The volume V of the test hole is determined by filling the hole with a material of known unit weight, in this case Ottawa sand. This sand is commercially available and usually sold in 50-lb bags. It is a sand of remarkably consistent texture, and when poured into a hole or other container in a standard fashion will assume a density of about 99 pcf. If you look carefully, you will see that the base of the cone has a lip that conforms to the hole in the plate. To conduct the test do the following:

1. Weigh the sand and jar, with cap (in advance).
2. Remove the cap and attach the cone to the jar.
3. Check that the valve on the cone is closed.
4. Invert the cone and place it on the plate rim. (Be sure that the rim is free of soil particles.)
5. Open the valve.
6. Close the valve when flow has ceased. (Note: This is, for some reason, fairly easy to forget.)
7. Disconnect the cone and reweigh the jar and cap (residue and jar).
8. Subtract (7) from (1) to obtain the weight of sand used to fill the hole and the cone.

Steps 1–8 will yield the field data needed to determine the volume of the hole, V. All other required data is normally determined in the laboratory.

This includes two items: the weight of the sand necessary to fill the cone and plate [this, subtracted from (8), yields the weight of sand in the hole], and the unit weight of the sand. The procedures for determining these items are simple, standard laboratory procedures and would probably be performed by a technician. While there are many details of the field and laboratory procedures, just remember what you are after and that you are simply filling a hole of unknown volume V with a material whose unit weight you have determined. Thus if the sand used to fill the hole weighs 3.30 lb and the density of the sand is determined to be (in fact) 99.0 pcf, the volume V is 0.0333 ft^3.

Schroeder (1980, p. 139) presents a complete set of data for the sand cone method. This test, and most others noted in this book have ASTM designations, and are described fully in ASTM publications.

When the test is completed in the field, it is a good idea to retrieve most of the spilled sand for reuse by shoveling it into a clean canvas bag, not so much because it is expensive, but rather to avoid running out, thus necessitating delays and repeated laboratory calibration work on a "new" sand. When you do this, avoid contaminating the sand with soil from the site.

Finally, the weighing and moisture content determination of the soil from the test hole should be done immediately after the sand weight is ascertained, to avoid errors through drying. This would be especially important on a hot, sunny day.

The Water Balloon Method

A more recent method for determining the volume of a density hole is through the use of a rubber balloon apparatus, where water is used to fill the hole. A thin latex rubber balloon within the hole contains the water and stretches to conform to the shape of the hole. Interestingly, in the earliest years of field-density testing, a board with a hole was placed over the test hole, a balloon was stuffed into the hole by hand, and water was poured into the balloon. The Ottawa Sand Cone Method apparently was developed because of the inherent inaccuracy and sloppiness of the original water balloon method. The new water balloon method, however, utilizes a device consisting of a graduated glass cylinder housed in an aluminum protective guard, with appurtenances allowing for quick, neat, and accurate determination of the volume of the test hole. The device is called Volumeasure and is available from Soiltest, Inc., Evanston, Ill. Other similar devices, such as the Washington Densometer, are used.

The devices were not available when I was doing fill control inspection, but I understand through talking with those who have used them that they are quicker than the sand cone method, each test taking about 20 minutes. The sand cone method I would judge takes perhaps twice as long but these estimates would be contingent upon the texture of the fill, particularly as it may affect the preparation (scraping) and digging of the hole and the unpredictable interference of gravel and cobbles. The water balloon method would seem to be inherently more accurate because of its use of water, which, of course, has an invariant density (for all practical purposes) and requires no calibration. Water is also cheaper than Ottawa Sand.

Jacked Sample Method

Having done site investigations involving test pit inspection and evaluation, I developed a means of extracting relatively undisturbed samples by jacking a section of scrap shelby tube (a hollow metal tube), about 6 in. long, into the wall of the test pit, using a 7-ton aluminum jack. Figures 8.4 and 8.5 illustrate the method. After jacking the tube into the wall, extraction is accomplished by digging the filled tube out of the wall with a geologist's pick or sturdy claw hammer. The samples thus obtained can be used for testing and analyses as with any undisturbed samples, including textural classifications (sieve and Atterberg limits), unconfined or triaxial compression testing, consolidation testing, and density–relative density determinations. This capability, in my opinion, more than compensates for the lack of blow count data that would routinely be available through conventional boring operations with a drilling rig.

For those who might use the technique for general foundation testing and analysis, the effects of anisotropy may be significant, inasmuch as the samples obtained will be approximately 90° out of the normal vertical orientation, and strength–deformation test results may be seriously affected accordingly. Textural and density test results will, of course, be unaffected. When deemed necessary, or when convenient, samples can also be obtained by jacking in a vertical direction, in which case, however, you will need something to jack against, like the bucket of a backhoe. The disadvantage here is that you'll need a backhoe and a backhoe operator to assist. Horizontal jacking can be done alone.

The jacking method described has been used only in test pit sampling of natural soils. Whether or not it will be easily extended to sampling compacted fills I am not certain, but I would expect that the texture of the fill and the energy of compaction would need to be considered. For example, an essentially granular TALB, containing significant quantities of coarse to medium gravel, compacted to a high percentage of modified Proctor, would be very difficult to sample by jacking. It would also, I might add, be very difficult to dig in (for the sand cone or water balloon method). These would be the blister densities that I referred to earlier.

Chunk Sample Method

As surprising as it may first seem, the very best type of sample one can use for general laboratory testing is a plain old chunk of the soil, assuming that reasonable care is taken in obtaining the chunk. A typical method of obtaining such a sample is to isolate a knob of the soil by carefully digging a circular trench; the chunk is then obtained by using a sharp spade to cut and lift out the chunk. Another method would be simply to carefully tear

FIGURE 8.4 Jack sample method.

FIGURE 8.5 Extracting a jacked sample from the wall of a test pit. (Whadd'ya mean, it's the wrong lot?)

or pull a sample from the wall of a test pit, avoiding a severe mashing or shearing action. The reason that such samples may be regarded as best is that the act of sampling does not significantly damage the sample. All other types of conventional sampling (e.g., standard penetration and shelby tube) require driving or pushing a sampler into the soil, producing disturbance by shock and/or side shear between the sampling device and the soil.

The moisture condition of a chunk sample can be preserved by wrapping the sample with a clinging type of clear food wrap. If necessary (for long storage), the sample can be further encased by smearing with melted wax.

The principal difficulty of strength–deformation testing of chunk samples is that one must trim the sample into some prismatic form, usually a cylinder. For density determinations, however, this is not necessary if one can devise a method for measuring the irregular volume of the chunk. One suggested method is to fashion a chunk-volume determinator. A large can with an overflow weephole and attached small-diameter graduated cylinder (or pipette) would suffice. One would fill the can with water to the level of the weephole, empty the pipette, and then carefully submerge the wrapped chunk sample into the can. The water displaced into the attached pipette would be a measure of the volume. The weight of the sample and its moisture content could then be determined in the usual manner, and the dry density computed. In doing this test, it would be important to wrap the chunk very carefully to ensure that there are no air bubbles of significant size trapped in or between the wrapping folds.

The Nuclear Method

Density determinations by nuclear methods might also be considered. While I have no first-hand experience with the use of the device in the field, I understand that the apparatus emits gamma radiation into the soil and that the measured rebound is calibrated to the density of the soil. In a similar way, the moisture content is determined by using a high-energy neutron source. As might be expected, these devices are very expensive. When calibrated and working properly, their major advantages are instant answers and speed of testing, thus enabling the performance of many tests in a short period of time. I understand that for highway work, a unit can be mounted on a jeep and test readings can be taken virtually on the run.

Their high cost limits their use, sensibly, to very large jobs or to organizations that routinely do major earthwork jobs, such as earth dams or highways. For smaller companies, which would probably not use the device day in and day out, I understand that there is a problem with maintaining the equipment in reliable working order; that is, the equipment tends to get out of calibration with disuse. Another problem is the bureaucratic involvement with the nuclear regulatory officials in Washington. My perception of these disadvantages stems from my only involvement with the device, which occurred about 20 years ago. A friend and colleague from a local geotechnical consulting firm gave the device (sans nuclear element) to the college for instructional purposes. That's when the letters from the AEC (Atomic Energy Commission) started coming. It turned out that merely owning the *container* that once housed the element required conformance with certain procedures.

Whether or not the newer models of the nuclear density gauge have similar calibration problems, I cannot say for certain. I would expect, however, that regulatory involvement is, if anything, more intense. My suggestion to those who would consider the device: Check it out first.

8.1.3   The Testing Program

After a review of the important features of the pressure bulb and the various ways of performing field-density tests, the questions that arise are: Which test method(s) should we use? How much field testing is enough? I have no pat answers, except to say that a lot of good judgment and experience will be needed to provide good answers.

The methods one uses will depend, among other things, upon the size of the job; the texture and homogeneity (or lack) of the borrow; the experience, expertise, and preferences of the controlling agency; and (to be blunt) the intelligence, experience, and even the personality of the inspector. Hopefully, the choice of methods and how to employ them will be aided by descriptions in the preceding section and elsewhere in this book. The second question, relating to decisions regarding how much testing to specify for a given job, is a much more complex issue. Included among the considerations, I would recommend the following:

1.   Cost.
2.   Consequences of distress or failure.
3.   Environmental factors.
4.   Degree of verification: expected versus warranted.

Before getting into some commentary on each of these items, it will help to set a perspective by introducing the notion of a *sampling ratio*. First, we can assert that it is clearly not possible to test *all* of the fill, so we have trapped the answer: somewhere between zero and all, and undoubtedly closer to zero. (That's not much help, but it's a start!)

In the fill job cited earlier for a major earth dike system, Hendron and Holish report one density test for each 2000 yards of fill, or approximately 1500 tests for the 3 million yards of dike. Assuming 6-in. diameter holes approximately of hemispherical shape, the total amount of soil tested was 2.42 yards, yielding a sampling ratio of 0.809×10^{-6} or, in round numbers, $1/10^6$.

Utilizing the pressure bulb concepts presented in Section 8.1.1, similar analyses could be made for any project, whether for soil sampling, destructive testing (field densities, as above), or nondestructive testing (plate bearing, for example).

I am not aware to what extent this kind of analysis has been done in professional practice, inasmuch as my career has been largely academic. What professional experience I have has been largely on small jobs where the number of borings, amount of testing, etc. is largely and more easily determined by the common sense and experience of senior engineers. It may be that the process of pursuing the elusive answer of "optimum sampling ratio" (if indeed one exists) is extensive in practice, but if so, the investigations and results do not appear to have been published yet, at least as far as my literature review has revealed. Such a "techonomic" analysis would be very interesting and valuable.

There are a number of reasons I perceive that might explain the dearth of information in this area, the first of which is that no one who is knowledgeable enough, and is privy to sufficient data, has thought of the idea of publishing an article or paper on the subject. Or it may be that they simply don't have the time. Clearly it would only be those people in full-time professional practice over an extended period who would possess both the knowledgeability and data to produce such a work. Unfortunately, it is only the academics who have the luxury (i.e., time) to produce scholarly papers. Indeed it is (or should be), next to teaching, their responsibility to do so. But many lack *any* experience (or interest), and almost all lack the data. Another reason for the lack of literature on this important topic might be that the data available within a professional organization is considered proprietary. Since it takes considerable time to collect, compile, and analyze such data, it is understandable that one would not wish to share it with competitors.

How may this void be filled? I see two possibilities. First, the data from public jobs, that is, those done by government organizations and therefore using tax revenues, should be available for the asking. The second avenue of approach would be an appeal by an appropriate organization for the voluntary contribution of the needed raw data. I would suggest that the Geotechnical Division of the American Society of Civil Engineers consider this solicitation. The compilation and analytical work could be done by academics with strong interests (and some experience) in the area, including professors and the graduate students whom they supervise on research projects and thesis work. To enhance the validity of the work, communication could be established between the contributing firm and the academics, including the participation of professionals on student (thesis) advisory committees. From recent first-hand experience, I can say that there is strong interest and movement toward greatly increased cooperation between industry and universities. This, it seems to me, would be an ideal way to fill the techonomic void that seems to exist in geotechnical engineering, while at the same time serving the broader purpose of industry–university interaction. It would appear to me that both would benefit. I further believe that this would be a legitimate activity for academics in that the work would be serving the needs of industry rather than competing, inasmuch as—if I am correct—industry does not have the time to pursue such applied research. Finally, I also believe that the economics of engineering practice is a badly neglected area of study and understanding among academics. From a recent conversation I had with an experienced consultant, I gather that there exists considerable lack of awareness of the effects of economics on design decisions among practicing engineers. Thus, the approach I suggest may have beneficial effects on both ends.

Cost Factors

Sowers (1979, p. 291) has reported that:

the cost of an adequate investigation (including laboratory testing and geotechnical engineering) has been found to be from 0.05 to 0.2% of the total cost of the entire structure. For critical structures or unusual site conditions, the percentage is somewhat higher, from 0.5 to 1%.

While I have not met or communicated with Professor Sowers, I was interested to note the dichotomy he makes between costs of geotechnical work for structures and critical structures, an approach similar to that which I adopted independently in devising my concept of a design flowchart (see Figure 1.2), where I refer to cheap jobs and expensive jobs. As described earlier, the former relates to jobs where the total cost is low and the consequences of distress are not severe, allowing one to design on the basis of the use of inexpensively obtained index properties, published charts, and experience. Expensive jobs were for structures (Sower's critical structures) where the consequences of distress or failure are of acute concern, thus sensibly prompting decisions for more expensive sampling and testing to determine engineering properties necessary to design a more foolproof foundation.

Buried among Sower's cost estimates is the fraction that might be allotted to fill control costs. So one approach to deciding how much density testing to do (i.e., choosing a sampling ratio) would be to base the decision, at least partly or initially, on budgetary considerations and then see how the answer conforms to technical judgments. This approach is of course nothing new to business people who often start out by asking, "What can we spend?" then working backwards to flesh out the line-item details.

It is interesting to note that Professor Sowers is designated in his book as Regents Professor of Civil Engineering, Georgia Institute of Technology; Senior Vice President and Consultant, Law Engineering Testing Company. In his preface, Professor Sowers also notes that there is "more than a century of geotechnical experience in the Sowers' family, spanning three generations." One can see that Professor Sowers is one of those rare individuals who somehow finds the time to both do the work and write about it, hence the valuable data he graciously shares with us on costs.

Consequences of Distress

In considering the multitude of decisions regarding costs and the concomitant details of foundation work for a structure, one probably subconsciously thinks about what would happen if the structure cracks, moves, or even fails completely. From a scientific or technological point of view, however, it would be much more beneficial to the design process if one *consciously* considers such factors, since the impacts can range from the trivial to the catastrophic in terms of costs measured in many ways. Tangible losses could include property and perhaps even lives, and intangible losses those of reputation and peace of mind. (If this seems overly dramatic, I remember that an engineer committed suicide after the collapse of the mountain wall of the Vaiont Reservoir in Italy, which resulted in the death of 2000 people in the early 1960s.)

Details of these considerations could include the following: What will be the use of the facility? As a mundane example of a compacted fill, consider a fill needed for a parking area for light trucks and automobiles at the rear of a warehouse, a job which I worked on many years ago. Our recommendations were to compact to 95% standard Proctor, install an asphalt (flexible) pavement, and grade for surface runoff away from the structure. We further stated that surface patching would be needed in the future.

In this example of a cheap job, without going into detailed cost studies, the project engineer decided that the cost of modified Proctor compaction and a high-quality pavement was not warranted and that considerable settlement was tolerable. For that reason, the flexible pavement was recommended, one that would conform with the settlements as they occurred. If too much local settlement occurred, resulting in puddle formation, inexpensive patching could rectify the problem. As I recall, the the amount of field density testing was zero, the control being accomplished by observation on site.

If the fill were to support a slab on grade, covered by a terrazo tile for the reception area of an office facility, the recommendations for compaction, testing, and field control would have been significantly more stringent and, consequently, more costly.

Environmental Factors

What about nuclear power plants? I am *not* suggesting that nuclear power plants be built on compacted fill! I *am* suggesting you read a book that deals with matters of risk, safety, and consequences in a readable and interesting way: *Of Acceptable Risks*, by William W. Lowrance, Harvard University. The book is short (180 pp.), scientific, but with almost no heavy mathematical treatment, and above all, timely. I believe it will be of interest not only to those involved in construction but also to anyone who is concerned about the burgeoning moral and technological complexities of our society.

With respect to the more prosaic aspects of consequences of distress and how they affect the testing program for fills, one must not fail to consider the applicable laws, building codes, and permits dealing with the construction, both from a safety standpoint (OSHA) and those dealing with potential environmental consequences, the latter becoming more and more prevalent in all areas of construction. Indeed, while I am not qualified to go into the subject, *hazardous* landfills have become a major national concern. Having lived and worked in urban New Jersey most of my life, I am more than normally sensitive to the problem, professionally and otherwise. In a survey conducted by the U.S. Environmental Protection Agency (August 1983), New Jersey was identified as having the largest number of sites containing hazardous waste materials (65 of 406). As a specific example of hazardous fills, emissions of radioactive radon gas from the soil surrounding several homes in a neighboring town have been detected recently. It seems that many years ago a company that made watches with luminous dials (radium) was active in the area. Speculation is that they disposed of their wastes on site, but that fill was later obtained from the abandoned site for grading of other residential lots. And, of course, everyone has heard about Love Canal.

I expect that within 10 years a book will appear in this series entitled *Avoiding Construction of and on Hazardous Fills*.

In terms of the testing program, I can only speculate that there will develop a need, indeed, enforced requirements, for more tests for toxic and hazardous substances, particularly in the more industrialized, urban areas and within haul distances of those areas. Then again, Times Beach, Missouri, no metropolis, was rendered uninhabitable by the spraying of dioxin-laced oil, done for the purpose of controlling road dust. (Dioxin is a deadly poison.)

Verification

Just as it is not possible to test all of the fill, neither is it possible to provide *absolute* verification of the compaction specifications, and clients, lawyers, and regulatory agencies should understand this.

There are other factors that can complicate the problem of verification, many having little to do with technical considerations, but seriously affecting the testing program and its cost. For example, it would normally be the project engineer on a job who would ultimately sign his or her name to the report or letter that certifies compliance with specifications. Typically, at least in my experience with small jobs, the project engineer would supervise field inspection work from the office, with only occasional visits to the site. As described earlier, present earthwork construction practice is such that the field inspector is very likely to be young or inexperienced. Hopefully, with the advent of civil engineering technology programs, and perhaps the future development of other technical support people, inspection work will be greatly improved by the creation of a group of people, whatever their titles, who will develop the skills and experience to provide the needed field control. I further suggest that such people will have to be compensated at higher levels as their skills and judgment improve with education and experience. The apparent costs of construction testing programs (i.e., fill control) will increase, but it is my belief that the actual costs will go down in the long run by virtue of the greater cost reductions related to reduced postconstruction damage and the attendant lawsuits. I believe the current practice of "kicking young inspectors upstairs" about the time they learn to perform well is self-defeating.

In any event, the degree of certification that will pertain to any job will be a significant factor in planning the testing program. There are, perhaps, three levels of certification that I can perceive:

1. That which is governed by the expectations of the client.
2. That which is legally required.
3. That which is dictated by your own standards of performance and reputation.

The client may be a company that has its own engineering staff, in which case communication can be established to ensure that there is an understanding on the part of both parties of what is intended and expected. If the client does not have engineering advice (internally), it is likely that he will depend exclusively on your professional reputation to do what is needed. In this case, I suggest that it still makes sense to communicate your intentions *and what it will cost* (at least in approximate terms) so that there will no future shocks or major surprises. Continuous communication, as inevitable changes occur, will also be beneficial to all concerned parties—trite perhaps, but no less true.

On some jobs, certification of construction operations like filling is legally required. Obviously, in such a case, the extent of field testing, and thus its cost, will undoubtedly increase. Case Study 2, Section 4.2.1, was such a job. Not only did the specifications call for supervision and certification, but it developed that the source of these requirements (particularly the certification) was a state regulating agency in Newark, N.J., to which I was required to send a letter certifying the fill. If you reread Case Study 2, you will see that this created a rather sticky situation. (To this day, I do not understand how a regulating agency in Newark got involved with a small construction job 40 miles away. Numerous other similar jobs in New Jersey which I have done did not have such a requirement. I didn't ask.)

Finally, if there are no legal requirements for certification, and the client is of the do-what-you-need-to persuasion, the degree to which you certify the fill is up to what might be called "company policy." And here it can get a little delicate, especially in view of our litigious society, a subject I touched on briefly earlier. For now we come to what lawyers call exculpatory clauses. These are statements that are inserted in reports, contracts, and plans that are designed to evade responsibility for anything that subsequently goes wrong. In legal parlance, they are also called hold harmless clauses.

I do not intend here to get into the legalities of the language of contracts, for I have little expertise or experience in that area, but I do wish to make some comments pertaining to the language contained in correspondence with clients, usually in the form of a cover letter accompanying a full report. The phrases that have always disturbed me are those which say, in effect, "We are not responsible for anything in this report." Call it exculpatory, call it hold harmless, I call it avoiding responsibility. Of course, such statements never appear in the plain language I give as an example. The wording is more subtle, typically very brief, and usually occurs in the very last paragraph (I suspect, so that the reader won't be as likely to notice). I have seen similar types of disclaimers stamped on drawings, very commonly on boring logs and soil profiles. My reaction when I see such phrases is, "What are we being paid for?" (as an engineering firm) or, with some sarcasm, "Does this mean we don't get a bill?" (from the drilling contractor).

I hasten to make an important distinction between generalized, all encompassing exculpatory statements and very legitimate statements defining limitations. An example of the latter is afforded by the following paragraphs, which were standard inclusions in all reports of the firm of Woodward–Clyde–Sherard and Associates.

Limitations—The recommendations made in this report are based on the assumption that the soil conditions do not deviate appreciably from those disclosed in the borings. If any variations or undesirable conditions are encountered during construction, the Soil Engineer should be notified so that supplemental recommendations can be made.

This report is issued with the understanding that it is the responsibility of the Owner or his representative to ensure that the information and recommendations contained herein are called to the attention of the Designers and incorporated into the plans, and that the necessary steps are taken to see that the Contractor and the Subcontractors carry out such recommendations in the field.

Notice the reasonableness of these limitations. Notice also that there is an implied inference that "we are confident in, and proud of, our professional competence and the correctness of our recommendations." In my view, that is a very important part of the successful practice of professional engineering: development of a feeling of confidence on the part of the client that the work for which he is paying is of the highest possible quality. It seems to me that one destroys this image by incongruously adding exculpatory statements of the kind described.

It may also be helpful to point out that exculpatory clauses do not always hold up in court. Richards (1976, p. 34) describes a case where the State of New York was "held liable for damages . . . despite exculpatory clauses in the contract." This was a situation where subsurface conditions were "vastly different than originally forecast on the plans" (supplied by the state).

Civil Engineering, the official magazine of ASCE, advises that "an owner cannot be assured of absolute protection by exculpatory clauses" (May1975).

With respect to a final report to a client, then, it would seem to me that some verification of the fill compact:on should be supplied, but with appropriate and reasonable limitations included, with perhaps specific reference to the sampling ratio and its relationship to the degree of verification that is possible. Carefully chosen language will protect the company both legally and with respect to its professionalism in dealings with clients.

Finally, one should recognize that, whatever the testing program, its cost, and its sampling ratio, it is the quality of the field testing work that is the most important determinant of whether or not the assessment of the fill is correct. It's like that most infallible system of computer programming, GIGO—Garbage In, Garbage Out. Better a bright, observant, experienced person with a pocket penetrometer, than a costly testing program of sophisticated tests conducted by an inspector with little sense of responsibility.

8.2   The Compleat Field Man

Before proceeding with a detailed description of field activities of the fill inspector, which I shall do with a case history approach, I would like to offer some commentary relating to equipment that a person who expects to be active in construction field inspection work can use as a check list in preparing for a field assignment. It is also intended and suggested that an employer (such as a geotechnical consulting firm) consider maintaining supplies of certain items so that any employee who is given a field assignment will be able to sign out the needed gear, with assurance that it will be complete and in working order. As I envision the process, it would be the responsibility of a staff person in the firm, perhaps a laboratory technician, to maintain the equipment in working order, conduct necessary inventories, and order replacement supplies and equipment as needed. With such procedures, the chance of a person discovering an important or vital piece of equipment missing upon reaching the site will be minimized.

Because I have been involved and interested in field work for all of my careeer, I have compiled notes on many aspects of field work, including general reconnaissance work, boring inspection, test pit inspection and sampling, load testing, and, of course, field-density testing. Rather than cull the information pertaining only to fill control work, I include it all in the hope that the information will be more broadly useful. Certainly, no field person will be so highly specialized as to be assigned only to fill inspection work, so a more general treatment makes sense.

With apologies to the increasing number of young ladies entering all phases of the work force, increasingly so in the technological areas, I have retained the title, "The Compleat Field Man." I refuse to get into the "his/

her" style of writing and talk about personholes. I believe there are certain limits beyond which the rewording of the English language is unjustified.

First, the list. Any item followed by an asterisk is explained more fully, for clarification, after the list. Items without asterisks are either self-explanatory or are described elsewhere in the book.

A. *Personal Gear*
   1. Clothing of appropriate capacity.*
   2. Foul-weather gear:
      (a) Rain: slicker, rainhat (hard hat), boots.
      (b) Cold: handwarmer, insulated boots, special socks, suspenders, wool watch cap or ski mask, writing gloves, ski mittens, chapstick, facial cream.*

B. *General Equipment*
   1. Distance measurements:
      (a) Knowledge of pace length.*
      (b) Metallic tape (50 ft).
      (c) Set of chaining pins.*
      (d) Stanley steel tape (retractable).
      (e) 6-in. pocket ruler (zero end = physical end).
   2. Elevation measurements*:
      (a) Folding rule (6 ft).
      (b) Locke level.*
      (c) Level rod (or facsimile).
   3. Direction measurements—compass.*
   4. Record keeping*:
      (a) Pocket notebook (plus protective plastic envelope).
      (b) Pen, pencil + sharpener *(no eraser)*.*
      (c) Camera, flash bulbs, film (fresh batteries).
      (d) Speed message forms.
      (e) Pocket cassette recorder (fresh batteries).
   5. Slope and plumb*:
      (a) Carpenter's level.
      (b) Plumb bob.
   6. Tool bag.

C. *Geotechnical*
   1. Penetrometer(s)—pocket.*
   2. Torvane.*
   3. Geostick.*

4. Sample bags (small, heavy plastic).*
5. Sample bags (large burlap).*

D. *Miscellaneous*
1. Hand sledge.
2. Stakes.
3. Keel (crayon).
4. Large shovel.*
5. Garden spade.*
6. Geologist's pick.*
7. Soft rock hammer.

E. *Special (as appropriate)*
1. Test pit sampling*:
   (a) Shelby tube sections (approximately 6 in.).
   (b) Lightweight hydraulic jack (7 tons).
   (c) Jacking kit (wood blocks, base, plate, tee).
2. Field-density testing*:
   (a) Scale (rugged, triple beam, counterweights).
   (b) Oven (or other device for moisture determination).
   (c) Sand cone apparatus (or equivalent, for hole volume).
   (d) Sieves (full size #4, full set of miniatures).*
3. Load testing*:
   (a) Engineer's level.
   (b) Aluminum cubes (approximately ½ in.).
   (c) Flashlight.
   (d) Hand telescope or binoculars.

Clarifying Comments

A.1.   By clothing of appropriate capacity I mean that which will accommodate all tools, devices, and record-keeping equipment that can, or should be, carried on the person. Thus, clothing with many and large pockets should be worn. For example, many people would not carry a notebook and pencil on their person, opting rather to leave writing materials in (say) the construction trailer. The trouble with that procedure is the strong likelihood of forgetting to write down some important observation that may have been seen down at the bottom of an excavation, for example, or at some other location remote from the trailer. Frequently, one would be distracted by some other matter, and by the time one reaches the trailer, the item will very likely have been forgotten. Sometimes, of course, the observation will include measurements as well, so the pad and pencil, *and* the measuring device(s) should be on the person. There is, of course, a limit to what one can comfortably and reasonably carry, and circumstances will usually dictate sensible choices. But certainly a pad, pencil, 6-in. ruler (maybe a Stanley tape), and possibly an inexpensive small camera, worn on the belt, would be carried at all times. Other equipment on the list that were deemed suitable in advance for the particular job would be in the trailer or in your car for use when needed. A camera is strongly recommended, whether carried on the person or not, in that very important events can be recorded more or less infallibly. Also, much time can be saved by eliminating sketching and writing that otherwise would have to be employed. Sketches and notes, moreover, are considerably more subjective!

A.2.b.   Cold weather gear is a necessity, not only for personal comfort (and perhaps frostbite prevention in *really* cold weather), but also to enable one to perform the field work, much of which involves a great deal of writing. Pile inspection is probably the best example in that driving records of the hammer blows must be recorded per foot of driven pile, and more often than that when the pile reaches its required degree of resistance (called "fetch-up"). Writing with nearly numb hands is no fun, and it gets to be virtually impossible to do so legibly. Hence, the handwarmer suggestion for the cold weather inspection job. There are pocket handwarmers available that utilize lighter fluid. You merely light the wick in the morning for a few minutes, blow it out, put it in a protective pouch, and (to my amazement) it stays hot all day, even in the coldest weather. There are also solid fuel types available.

The other cold weather gear cited is intended to keep the other extremities warm, or at least reasonably so. Outdoors experts recommend silk socks under heavy wool socks, the effect being a trapped layer of warm air for additional insulation. If you are especially sensitive to cold, you might even consider electrically heated socks, particularly if you have a long-term cold weather assignment and the nature of the job requires more standing than moving, such as pile inspection. I have found the difference between standing and moving jobs to be dramatic in terms of keeping the feet warm. Especially avoid tight footwear. A boot one size larger than one's street shoes may be helpful to accommodate the heavier socks.

While I have never confirmed it personally, I recently was told by a construction worker that suspenders, rather than a belt, are more conducive to warmth. This makes sense, since the entrapped warm air would serve as added insulation, as with the layered socks. Layered, loose clothing is, in general, the better choice.

The watch cap is important for two reasons. First, it directly protects the ears, and second it helps keep your feet warm. Say it again, you say! Listen: If the body is regarded as a heat sink, when heat is lost, where will the painful sensations be felt first and most intensely? The scientific answer, at least from a heat loss standpoint, is where the area–volume ratio is highest. And where is that? The ears, the fingers, and—of course—the toes. So wearing a hat keeps your feet warmer. (Dr. Gabe Mirkin, fitness expert for WCBS radio, talking about running in cold weather, reported that at 20°F, 40% of the body's heat loss would be through an uncovered head; at 5°F, 70%.)

The ski mask may be advisable for extremely cold windy weather, when frostbite is a distinct possibility. For extended cold weather assignments, a beard makes good sense. Ladies and clean-shaven men can use facial cream.

Writing gloves refers to a pair of gloves that will provide some protection but still allow you to write legibly. These might be helpful on jobs where almost continuous writing is required, as with driving logs. Leather ski mittens could be worn over the finger gloves when writing is not required. (When your mom sent you out with mittens, she knew all about Monahan's area–volume theory.)

B.1.a.   A field person should determine his or her pace length. A simple way to do so would be to lay out a measured distance of 100 ft with a metallic tape, and simply count the number of paces to cover that distance. Do not exaggerate the individual steps, as football referees do, but rather walk at a comfortable gait. It is not only a matter of comfort, but also one of accuracy, in that you'll often need to determine fairly long distances in the field and striding unnaturally will be both tiring and inaccurate.

B.1.c.   When a distance measurement should be more accurate than pacing would seem to permit, a 50-ft metallic tape and a set of chaining pins may be used. The chaining pins permit a person to tape a distance alone, since metallic tapes have a rectangular metal loop at the zero end. You proceed as follows: Pin the loop into the ground, go forward 50 ft, and mark the point. Then pull the rear pin forward as you would a fishing line. Chaining pin sets contain two rings that may be attached to one's belt (or suspenders!). One starts with all pins on one ring. As the 50-ft lengths are measured, the pins are transferred to the other ring. In that way, it is simple to keep an accurate count on the number of 50-ft lengths measured. This technique is helpful because one often needs to measure distances without assistance.

B.2.   For appropriate elevation measurements (or differences in elevation), one can use a Locke hand level and either a 6-ft folding rule, level rod, or facsimile. If there is no one to assist, the opened 6-ft rule can be jammed or otherwise buried in the soil sufficiently to stand alone, and leveling along a line between two points can be done in the usual process of alternate backsights and foresights, using the Locke level as a substitute for the more usual engineer's level. For those unaware, the Locke level is a hand-held tubular device, about 8 in. long, which enables one to sight along a horizontal line to a level rod at the backsight and foresight locations. The procedure using the folding rule is a bit awkward, and it would be preferable to use a more suitable level rod and the assistance of a second person as rodman. However, there may be instances where one does not have the luxury of an assistant. In fill placement work, where I have used the technique, it is probably only necessary to determine elevations periodically to within 0.5 ft to keep track of the progress of the work for purposes of interim reports, and the Locke method, done with reasonable care, will serve that purpose. Final grades, of course, would be determined by more accurate, standard surveying methods, and most likely would be done by others (i.e., professional surveyors).

B.3.   A small hand compass can be used to establish approximate directions. It also may be useful for locating shallow (buried) metal objects such as surveying markers or utilities, as long as one knows in advance their approximate locations by ties. A Brunton pocket transit can also be very useful.

For those not familiar with basic surveying techniques and tools, see any elementary surveying text. One of the very best (still), in my opinion, is *Elementary Surveying*, Vol. I, by Breed and Hosmer, Wiley (1945). There are at least eight editions, with copyright dates starting in 1906! There are, of course, many new and extremely sophisticated surveying instruments and techniques now available, like laser beam technology, but for the kind of stuff I'm describing here, refer to Breed and Hosmer.

B.4.   A few additional words about record keeping in general is warranted. As mentioned under "Clothing," one should always carry certain equipment on one's person at all times while engaged in field work of any sort. Note under B.4.b that I suggest and emphasize *no eraser*. This is because of the possibility that construction field notes or data may wind up as evidence in depositions or in court cases. Clearly, then, any erasures are suspect. It is good practice to avoid their use. When you inadvertently write down an incorrect word or number, simply and neatly cross out the error and add the correction. After a while, this procedure becomes a habit, and the value of your records is enhanced.

A second suggestion is to avoid ambiguous entries. On fill jobs, for example, use the letters "C" and "F" for cuts and fills, rather than " − " and " + " (requiring an interpretation). For approximate directions use "easterly" and "westerly," rather than "left" and "right." (On a highway location study, I once worked two weeks in the office plotting surveyor's topographic notes and preparing contour maps. His notes read "left" and "right" of centerline. He was out of the country on vacation, so we could not contact him for clarification of the ambiguous notes. Upon consultation with the project engineer, we agreed to guess which meant. We guessed wrong, and two weeks of work produced upside down topography!)

One way of avoiding ambiguity is to adopt the habit of explaining symbols used with a note explaining the symbol's physical significance, for example " + means the volume is increasing." Another helpful technique is to assume that another person will have to interpret your notes or data without any opportunity to question you. If you consciously take this approach on a regular basis, after a while it will become habitual to you, and the result will be consistently clear, unambiguous notes or data. I have observed over many years of teaching laboratory courses, including surveying, that students develop bad habits in record keeping, because in their academic setting, they will be interpreting their own notes or data, usually within hours or days, and so they subconsciously rely on their memory for interpretation.

Speed message forms are those commercially available snap-off forms with built-in carbons. They are very useful and time saving when you wish to mail short but important notes from the field and wish to retain a copy. The plastic envelope suggested for the pocket notebook is to protect against obliteration of important notes by perspiration or rain.

B.5. Many times in the performance of an assortment of field activities, especially inspection work, it will be necessary to measure horizontality or verticality, sometimes rather precisely. A 6-in. rule and a good-size, quality carpenter's level (one about 4 ft long is suggested) will enable one to measure quickly for horizontality. Incidentally, the measurement will be quicker, easier, and much less susceptible to error if the zero end of the rule is *physically* the end of the rule. On many pocket rules, this is not the case. A plumb bob and 6-in. rule can also be used to check plumb. The plumb bob has the advantage that it can be readily carried on the person.

C and D. The penetrometer, Torvane, and geostick have been described briefly in Section 8.1.1.

Sample bags of various sizes and certain digging tools will be needed for a variety of purposes. The small, heavy plastic bags might be used for holding small samples scraped from the wall of a test pit for textural classifications. The large burlap bag and larger shovel would be needed to obtain a borrow sample for laboratory compaction testing. A garden spade would be suitable for digging field density holes or for obtaining small bag samples. The geologist's pick would be useful in many ways, with one specific use being the retrieval of jacked samples from the wall of a test pit (see Section 8.1.2).

E.1. Test pit sampling has been described briefly in Section 8.1.2 and illustrated in Figures 8.4 and 8.5. The jacking kit listed is simply a wooden box containing the jack, wood blocks, base, a plastic plate, and tee. The tee is an aluminum, T-shaped part, the vertical section of which transfers the load from the jack piston to the hollow section of shelby tube. The horizontal portions support the tube (front) and the jack (back). The base is simply a wooden trough extending approximately across the width of the test pit, upon which everything rests. The wood blocks, of assorted arbitrary sizes are used as chocks to transmit the load from the rear wall of the test pit to the shelby tube section. The tube is thus jacked into the wall, periodically resetting the jack piston, changing blocks as needed, until the tube is fully jacked into the wall. The filled tube is then extracted by digging around its exterior with the geologist's pick. The entire kit is fairly heavy, but portable.

E.2. Field-density testing has of course been described extensively in prior sections. Detailed itemization of equipment has been omitted, however, because several options are available for the various operations. Note that under sieves, I list a full-size #4, and a full set of miniatures. The former is normally needed to determine the percentage of soil from the density hole which is retained on the #4 sieve, this quantity being needed to allow an empirical correction to the target value density. Recall that, because of the mold size, the +#4 material is excluded from the laboratory test. If some larger size is excluded, utilizing a larger mold, then that sieve should be brought to the field. For example, ASTM permits ¾-in. particle sizes if a 6-in. diameter mold is used.

The full set of miniature sieves (about 3 in. in diameter) would enable one to do a sieve test in the field by hand shaking. Such testing is not normally done as a matter of routine, but it can't hurt to have that capability in the field, especially if the sieves are available and not needed elsewhere. Also, there are some circumstances when field sieving might have an important purpose. One would be an attempt to test the idea of the compaction data book that I proposed and described in Chapter 4, as a means of dealing with the problem of changing borrow. Plots of grain size distributions are needed to apply the method, and hand sieving would provide more accurate plots than reliance upon visual estimates of gradation. Another circumstance might be one where the field person is very inexeperienced and has not yet developed the skill of visual soil classification. In such a case, hand sieving could be used as a check (or substitution) for visual classifications. (For those unaware, with practice it is possible to prepare fairly accurate grain-size distribution curves by visual and tactile inspection, utilizing techniques developed by Donald Burmister of Columbia University.)

It should be recognized that hand sieving in the field will introduce some error, inasmuch as it will not likely be possible to dry out and wash the sample in the sieving process, as is done in the laboratory. Without washing, fines will adhere to the coarser particles, thus *always* yielding erroneously low percentages of fines. If the quantity of fines is very large, the error may be significant. (It occurs to me that this problem would make still another simple applied research investigation to be added to those described in Chapter 5. The goal would be to develop a prediction capability for adhered fines corrections, to improve the reliability of the process used for the suggested compaction data book.)

E.3.   Load testing in the field is often a very important, time-consuming, expensive activity. A single pile load test, for example, can take a day to set up, several days to run, and cost perhaps $5000. Plate load tests can be relatively simple, 5-minute tests such as that described by Giddings (Section 7.1.1), or larger tests comparable in elaborateness and cost to pile tests.

It is beyond the scope and focus of this book to include details of large-scale load testing, but I offer some brief commentary about the role of the field person.

First, pile tests and plate bearing tests may both be thought of as plate tests because the load applied to the pile is transferred to the butt end of the driven pile by a thick steel plate, usually about 3 in. thick. Second, both tests measure loads and corresponding deformations. Loads are usually applied by a hydraulic jack, and a common method involves jacking against a dead load composed of tons of steel ingots piled on a reaction beam.

The plate is instrumented in several ways to allow independent means of measuring the movement of the plate, three of which are (1) extensometers (Ames dials, typically) mounted at corners of the plate, (2) engineer's level readings (foresight rods attached to the plate, backsight rods on selected immovable benchmarks on site), and (3) piano wire strung in front of a steel rule mounted on the side of the plate.

It is commonly the function of the field inspector to approve the setup, and participate in the running of the test, principally by reading the instruments that measure plate movement and, of course, recording all results.

The setup is done by the construction workers. This can take several hours, using cranes and the like. Details will not be included here, except to say that when they are done, you'll be asked, "Is it O.K.?" This can be rather sticky, for if you simply say "Yes," there's an inference that anything that goes wrong later may be blamed on you, with a "He said it was O.K.!" chorus from the contractor's people. To make matters more complicated, there is a good likelihood that the entire setup was done without your being present to observe the process. I recommend that, if at all possible, you make every effort to observe the process, at least intermittently, for it is much easier to spot and correct problems during the setup than after completion. For example, one item that should be checked, using the carpenter's level, is that the the reaction beam is horizontal in at least two directions. A typical limitation on the allowable slope is ½%. (A code will likely govern this.) Clearly, it is much easier to correct an unacceptable slope early in the setup (by having shims inserted) than when the full dead load has been added. (I participated in one pile test where the dead load exceeded 350 tons.)

Some of the equipment that is used in the test itself, or which can be helpful in taking the necessary measurements, is listed under E.3. The items all pertain to the measurement of the movement of the plate under the influence of the load increments and decrements. Generally, increments (decrements) are added (removed) when plate movement has ceased under the influence of any given load. The test is run continuously, so inspectors work in shifts. A failed test is commonly defined by some limiting net movement of the plate (i.e., not including elastic deformation). In New York City, for example, a net movement of a steel pile exceeding ¾ in. defines a failed pile.

The instrumentation apparatus is commonly supplied by the contractor and is part of the setup. This might include the Ames dials, rods, piano wire, and the means for setting them up in a firm manner, that is, in such a way that they will not be physically disturbed during the test. (These are some of the other things which should be checked before saying that meaningful "Yes.")

E.3.a.   In my experience, the agency that supplies the engineer's level and is responsible for its adjustment (and perhaps certifying that adjustment), is often left up in the air. I don't think it should be, and I recommend that this be settled in advance. What happens on a typical load test is that none of the independent sets of data agree, and difficulties (and even arguments) develop over which is more correct. With two foresight rods, a set of Ames dials, and a piano wire, there will be *four* sets of independent data. If the level is in proper adjustment, and is so certified in advance, the sources of discrepancy will be reduced. (Notice I say reduced; they're never eliminated.) Breed and Hosmer describe simple level adjustments, but often it will be necessary to send the instrument to experts for shop adjustments, especially since instruments used around construction projects are more likely to have been knocked around. An alternative, of course, is to get another instrument. It seems to me foolish to risk messing up an expensive test by using a level badly out of adjustment.

E.3.b.   Ames dials are usually mounted on three or four symmetrically located points above the plate, with their plungers initially pushed in. An arbitrarily set initial reading for a dial with a 1-in. range might be 0.026. As the test progresses, at some time late in the test (higher load), the plate will have moved to near the range of the dial, at which time a reset of the dials must be done—before the plate leaves the dials. Neat aluminum blocks are recommended for this purpose. One would simply write "reset dial" on the data sheet, read and record the dial (say 0.963), push the plunger upward, insert the aluminum spacer cube, and read and record the equivalent new dial reading (say 0.483). Alongside the "reset dial" notation, write $0.963 \equiv 0.483$.

An alternative to this procedure would be to remove the entire dial and remount it at the new setting, but this is more cumbersome in that you'll need pliers or other tools, and there is greater risk of otherwise messing up the instrumentation. Remember, it may be dark, cold, and raining or snowing, and you will be typically working in tight quarters and in close proximity to all the other instrumentation. (If you hear a "spronggg," you just blew the piano wire system. Of course, only you and the Lord will know!)

E.3.c.   A flashlight with fresh batteries will be needed if you are assigned a shift encompassing darkness.

E.3.d.   Admittedly, it may be superfluous, but a hand telescope or binoculars may be helpful to give you remote capability to read the dials. In that event, a rather powerful flashlight will be needed for night work. One caution, however: The pointers on some dials have a tendency to get stuck on the inner face of the glass, requiring a judicious tap or two with the eraser end of a pencil to unstick the pointer to get the correct reading; this is apparently caused by the accumulation of static electricity. You should check whether this occurs with the dials you are using. If it does, remote reading will not be advisable.

The idea of remote readings is not a matter of comfort and convenience, but rather one of a further precaution against disturbing the instrumentation. As noted, the test area is usually pretty tight and cluttered, and heavy rain can create a quagmire or even standing pools of water to further complicate things. Incidentally, this is one of the other things that can be checked if you have the opportunity to observe the "dock wallopers" setting up the test: the room for movement provided in the immediate area of the jack and instrumentation and the grading (topography) of the area. By tactful requests or suggestions you may be able to make your job easier later. I would not recommend that you demand or in other ways appear to be supervising, for you may inadvertently create a potential for later culpability should something go wrong, and, as Murphy says, it will. Also, the "dock wallopers" (the heavy equipment operators who set up the test) may resent it if you come on too strong.

8.2.1 Summary Comments

The list presented at the beginning of this section is based only on my experience, so it is by no means intended to be a complete treatment of the "The Compleat Field Man." But I think that the approach I describe is a sound one. I urge organizations to add to the list or to prepare their own, and to provide their field personnel, in a systematic way similar to that described, with the equipment and other backup that will maximize their potential for effective field performance. I suggest that individuals, particularly the younger, inexperienced field person, keep careful notes in the field, and regularly review and summarize their notes at the completion of a field assignment, with an eye toward improving performance on a similar future assignment. With apologies to Alexander Pope, "to err is human, but to screw up the same thing repeatedly is unforgivable."

8.3 Case Histories

To cover the remaining items I consider important in fill control field techniques and inspection, I decided to intersperse most of the commentary within descriptions of case histories, principally the one designated as Case Study 1, Section 4.1.1.

Rather than repeat myself, I suggest that you reread Section 4.1.1 to refresh your memory regarding some of the essential features of the job, notably those dealing with the focus of Chapter 4: major problems in fill control.

An approximate site plan and profile (prepared from memory) is shown in Figure 8.6. The job, while very small in size compared to some described in Chapter 7, was the largest I have been assigned to or involved with (about 150,000 yards). With the possible exception of a short job assignment in northern New Jersey, which I will describe briefly later, it was probably the

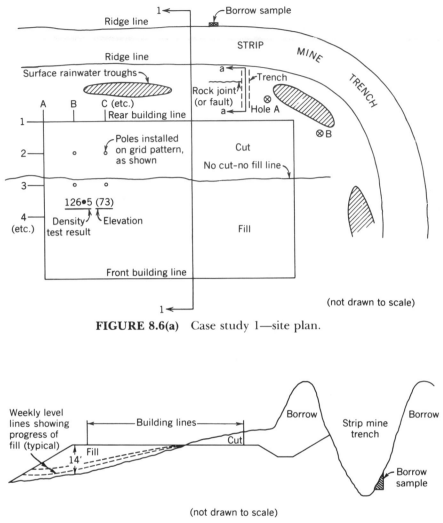

FIGURE 8.6(a)   Case study 1—site plan.

FIGURE 8.6(b)   Case study 1—profile.

most problem-plagued job with which I've ever been associated. I hasten to add that this was not through any fault of mine or the company for which I worked. Indeed, it was through our efforts that the job was straightened out and "completed." (The use of the quotation marks will be explained in due course.) Interestingly, it has been my perception that one learns a lot more from these types of jobs than from those which run relatively smoothly. I hope I am right, and that you will agree.

The job was complicated administratively at the outset by an unusual layering of interested parties, due primarily to its location in Appalachia, an economically depressed area. Because of the very high unemployment rate in the area, caused largely by the demise of the coal mining industry, incentives were developed to attract other industries to the area. One such incentive was site preparation work at no cost to the prospective industrial client. That was the case on this job. Thus, to start with, there were three levels of government involved: federal, state, and local. Add to that the architect, design engineers, earthwork contractor, soils engineers, and, late in the job, even geologists. The administrative complications were considerable. Even though I would be involved only with the fill inspection, and not with executive-level policy discussions, as the only representative of my company on site, I found myself making lists of the people involved in order to provide coherent reports to those to whom I was accountable. The list included affiliations and phone numbers. (This is a good early step in the record-keeping process on any job.)

As related earlier, I arrived in the late morning at the design engineer's office, and visited the site that afternoon. By evening I had become aware of quite a number of technical problems caused largely by the fact that the design engineers had called in our company much later than would have been ideal. (See Section 4.1.1 for summary.)

Before continuing with a description of the field problems on this case, and how they were resolved, you should be aware that I was in frequent telephone contact with the project engineer, at his insistence. While I was older than most inspectors would be, having been in the armed forces for 4 years, I was not much more than a geotechnical rookie, this being only my second summer's employment with a soils consulting firm. I did have earlier summer engineering jobs, but those were more broadly varied, including structural design and highway engineering. I had not yet obtained my PE license, being one year short of the experience required to take the test. The reason I include this background information is because it is typical for field inspectors to be relatively inexperienced. Thus, at least for small jobs, supervision is commonly provided by phone and, where possible, by short site visits by the project engineer. In many cases, then, the solutions to field problems are often augmented by the advice and instructions provided from the main office. However, there is a reasonable limit to how often one can or should call for advice. Deciding when to call, and with regard to which problems, is a problem in itself! In a nutshell, while some help is available, more often than not, you'll have to sink or swim by yourself.

Since filling had already started, one of the first problems I attempted to resolve was the obvious inadequacy of compaction capability, as reflected by the presence of a single compaction roller, shown in Figure 8.7. This photograph illustrates two problems written about earlier: inadequate compaction energy and a major imbalance between compaction and hauling–placement capabilities.

I first spoke with the earthwork contractor's superintendent about the two problems but shortly realized the best follow-up would be to prove the

FIGURE 8.7   Compactor, *circa* 1930.

inadequacy. This was done by my first field-density test. On my suggestion, we both agreed to observe the placement and rolling process. The fill was brought in by the scraper (see Figure 8.8) and placed to a lift thickness agreed upon, about 12 in. This was measured and then rolled with (as I recall) five passes of the roller. Using the sand cone apparatus (Figure 8.3), I determined the dry density to be significantly less than the 95% *standard* Proctor called for in the specifications. (Note the emphasis. At this time, the major problem had not yet been discovered.)

The corollary problem of equipment imbalance, while not susceptible to a proof as explicit as the roller energy inadequacy, was fairly easy to resolve, since the superintendent was already convinced that he needed heavier compactors. Incidentally, one of the things I learned from this and similar experiences is the almost extraordinary ignorance of many earthwork contractors (or their employees) about field density testing.* As you will see by the extensive amount of heavy equipment that subsequently was brought to the job, this was a sizable contracting company. The conclusion I have drawn from the puzzled reactions of contractors to the appearance of field density apparatus on the site, on more than one occasion, is that contractors have an "imbalance" themselves, in that they are heavily preoccupied with moving soil, but don't think too much about compacting it, perhaps an intangible major problem in itself.

---

*I do not use the term "ignorance" with any insulting or derogatory intent, making an important distinction between ignorance and stupidity. Thus, a very intelligent person can be ignorant of some fact or technique, as is the case and connotation here.

FIGURE 8.8   A scraper or pan.

After being convinced of the need, the contractor brought in a large number of Euclid trucks from another job that had been completed nearby. With some discussion and consultation all around, an idea evolved to use loaded Euclid trucks with overinflated tires for compactors. As I recall, the contractor estimated that he could safely inflate the tires to about 70 psi. Figure 8.9 shows two such trucks in the process of compacting the soil. Although not a standard procedure, it worked well. One of the things that had to be monitored fairly often and carefully was the compaction of the "missed" portion between the rear wheels. The truck operators were instructed to move laterally the width of two tires (about 4 ft) after the required number of passes had been completed.

The trucks were also used to haul and place soil as shown in Figure 8.10. Power shovels, not shown, were used to load the soil in the trucks. Notice that the Euclid is in the process of dumping the fill, and that a dozer is required for spreading. Unlike the scraper, the trucks cannot be used effectively for placing the fill in a lift of uniform thickness. The dozer operation is an added cost. Also shown in the photograph is a sprinkler truck (background, behind rear of the Euclid). The time period of this job was rather dry and the borrow got unmanageably dry, requiring the added steps of sprinkling and mixing before compacting. Also shown in this view are scrapers.

After addressing the immediate problems of compaction and equipment imbalance, necessary to keep the job moving satisfactorily, the problem of locating adequate borrow of acceptable texture was next. As reported in Chapter 4, this problem had apparently not even been considered before

FIGURE 8.9 Euclid trucks as compactors.

FIGURE 8.10 General view of filling operations.

our company was contacted. The soil for the filling that was being done prior to my arrival was obtained from the cut area toward the rear of the site (see Figure 8.6). As it developed, and was fairly evident from the beginning, the amount of fill required far exceeded the cut volume. First, the soil near the surface in the rear contained substantial amounts of organic material, including some rather large branches. Most of the latter were removed by hand and burned periodically, and some additional stripping and wasting was necessary. Second, we hit rock at fairly shallow depths in the cut area.

Careful earlier planning would have included a search for suitable borrow, including studies of regional soil maps (what I call paper reconnaissance), physical reconnaissance, negotiations with the owner of a potential borrow, some physical investigation such as test pits or borings to determine the extent and texture of the borrow, and finally the actual steps toward acquisition.

For several reasons, this sequence could not be followed. Most obviously, the job was underway and it was impractical to put everything and everyone on hold, especially because of the added factor of the unusual administrative complexities of the project. Also, as I understood it, available funding was tight or questionable. The upshot of it was that, after some rather quick local reconnaissance and phone calls, a decision was made to use the soil from a contiguous strip mine area for fill.

Figure 8.6 shows the location of the strip mines that were used for borrow. Strip mining is a procedure where the ore (in this case, coal) is at such shallow depths that it is more economical to strip away the overlying soil and then mine the soil in an open pit operation. Deep veins are, of course, mined from vaults reached by shafts and tunnels.

Access to the trough of the strip mine was found during reconnaissance through a pass several hundred yards away. With the cooperation of the contractor's superintendent, I traveled to the area on a bulldozer and had the operator cut into the base of the strip mine to expose a section judged to be representative of the soil. Figure 8.11 shows the exposed wall. As may be judged from the photo, the composition of the soil was extremely variable, containing many obvious small boulders and cobbles but also containing substantial quantities of all smaller sizes and categories: gravel, sand, silt, and clay. The high percentage of clay is responsible for the free-standing vertical face of the cut. Upon closer inspection and subsequent laboratory analysis, it turned out that the clay content was as much as 40% and was of low to medium plasticity. For most purposes, such a high clay content would be considered a poor characteristic for a structural fill. However, because of reasons already cited, the decision was made, with some reservations, to use the borrow. The advantage of proximity and easy availability, without hassle or acquisition problems (not to mention low cost), undoubtedly also contributed to the decision.

The borrow was of glacial origin, a till deposited directly by the glacier.

This explains the unsorted nature of the soil, resulting in the very wide range of sizes. The natural till, located beneath the excavated material from the strip mining operations, was very dense—a result of the large pressures exerted by the ice sheets that deposited the soil. The combination of co-hesiveness and high density resulted in a soil which was difficult to excavate. A practical ramification was that scrapers had to be pushed by bulldozers during construction excavation.

A sample of the soil was cut from the face of the borrow area, as shown in Figure 8.11, and was brought to the laboratory for compaction testing and textural classification. The soil was cut from a vertical section of the freshly exposed face in order to obtain a sample that was as representative of the borrow as was reasonably possible to obtain. The sample weighed about 175 lb, probably somewhat more than necessary, but it's probably a good idea to be generous and get more rather than less.

Shortly after the problems of immediacy had been resolved, the real bombshell occurred. An engineer from the industrial tenant for whom the site was being prepared arrived at the site, having been assigned by his boss to visit the site to see how things were progressing. What transpired is described in Chapter 4, Section 4.1.1.

After the specification blunder was rectified by changing the target value to 95% *modified* Proctor, the job progressed fairly smoothly for a while. As things settled to relative normalcy, I endeavored to establish methods control

FIGURE 8.11   Exposed wall of borrow area.

criteria with the contractor by suggesting to him that we run a test strip "field experiment." By this time, we had the results of the modified Proctor test on the borrow, and so a target value density (95%) was known. However, I explained to the contractor that it would be advantageous to both of us if we could establish a lift thickness and number of passes (with the loaded Euclid trucks) that would give us the target value density. Being a reasonable man, he agreed, and we laid out a strip (on the site so it wouldn't be extra work), measured the lift thickness as 12 in. and both observed and counted three, five, and seven passes over three segments of the strip. I then did density tests in each of the three segments. (One of the tests is shown in Figure 8.3.) Upon completing the weighings and computations, I reviewed the results with the contractor. It turned out that five passes were not quite enough, and seven produced a density comfortably above the target value. Thus, we rather amicably agreed that six passes would be the typical method. I could now exercise some degree of control other than the density test. This did not become an official part of the specifications, but rather a reasonable agreed criterion. I emphasized to the contractor that the six passes would only be valid if no other conditions changed significantly: borrow texture, borrow moisture, compactor, and lift thickness. If it was judged that one or more factors did change significantly, new methods would have to be established, perhaps by another test strip experiment *and* maybe even another laboratory compaction test.

Changes in borrow texture, and suggestions for how to handle this problem were discussed in Section 4.3. Moisture problems and suggested solutions were described in Section 7.1.1, under weather restrictions. Compactors and the effects of compaction energy have been treated throughout. Suggested compactor types and lift thicknesses are listed in Table 6.1.

8.3.1   The Hole

About the time everything seemed to be running smoothly, an incident occurred that made me a firmer believer in both Murphy's law and Monahan's theory of the perversity of inanimate objects. A worker came running toward me exclaiming rather excitedly, "There's a hole, a hole. . .," pointing toward the rear of the site. I strolled to the area, and sure enough, there was a hole. I photographed the hole (Figure 8.12), thinking: They'll never believe this at the office. The approximate location of the hole is shown on the site plan, Figure 8.6, hole A. (Another hole appeared a day or two later, also shown on Figure 8.6.) The 6-in. scale in the photo (always a helpful device for photo records) indicates that the hole was approximately 1½ ft in diameter. It was also rather deep, as evidenced by the fact that one could not distinguish the bottom visually, and a small boulder thrown in gave audible confirmation.

Because of the obvious concern that subterranean mining had also occurred at the site at some earlier time, or possibly that natural limestone

FIGURE 8.12   The hole.

cavities were the cause of the sinkhole, a decision was made to bring in a geologist to investigate. A brief description of the sequence of steps in the investigation, and the results, conclusions, and recommendations follows. Understand that I was not involved in any official capacity in this work. In fact, the filling operations continued, so I was otherwise occupied most of the time with my own duties. I must say that I did observe the proceedings with much interest, and, in fact, did a little surreptitious investigation and photographic record keeping on my own, mostly during lunch breaks, evenings, and weekends when no one was about. I did not confer at all with the geologist, after initially detecting an unmistakable disinterest to do so on his part, an unfortunate but not uncommon attitude of geologists toward engineers.

The first step was to bring in a dragline to excavate at hole A. In a few days a large hole was opened to a depth of about 25 ft, where a cavity was found with the unmistakable remnants of thick timber shoring. (Remember that no borings had been taken for this job, and I would surmise that old mining maps and records had been lost; either that or no one thought to search whatever records were available.)

During this time period, we got some rain and an interesting pattern of surface rainwater troughs formed, distinctly elongated and roughly paralleling the strip mine trench. This was definitive evidence of a pattern of surface subsidence, prompting (I surmise) the next logical step in the in-

FIGURE 8.13   The trench—a view of section a–a, Figure 8.6*a*.

vestigation sequence: a trench dug in a direction approximately perpendicular to the elongated subsidence troughs. A power shovel was brought in and such a trench was excavated, as shown schematically in Figure 8.6*a* and in Figure 8.13 (see the section a–a, 8.6*a*). As seen in the photo, a very distinct joint or fault was exposed. If you look closely at the top of the photo, you can see the third dimension of the joint on the surface. This surface crack was traced for 40 or 50 ft in a direction parallel to the troughs, the rear building line, and the strip mine trench and ridges.

With our fears mounting, the next step was to initiate a boring investigation to attempt to determine the extent of the problem. Figure 8.14 illustrates that this was done while the filling operation continued. The boring investigation (actually corings, as they are called in drilling in rock) was, of course, quite a logical next step. There had been almost no geotechnical investigation done before the filling operation had started.

Figure 8.15 shows one of the coring boxes, with the typical arrangement of the cored material from top to bottom. Note the shattered, broken nature of the rock nearer the surface caused by weathering. Most importantly, note the printed comment, "Void: 23.5 to 26.0."

As noted, I was not involved in any way with the investigation sequence and neither was the company for which I worked. Thus, I was not privy to all of the results, for example, of the coring operation. I heard informally

FIGURE 8.14 Coring and filling operations.

that an initial decision and recommendation was to relocate the proposed structure a few hundred feet away, but I was told later that the structure was, in fact, built as originally planned. I did not get any final, definitive word on what actually was done, but I'm sure curious, even to this day. I hope some day to return to the site, perhaps incognito, to look around.

The reasons I chose to include the descriptions of this section, fully recognizing that they have nothing to do with filling operations, is that I think the material is interesting and instructive to field persons, particularly the younger, inexperienced ones, to whom this chapter is particularly directed. I think it illustrates the importance of field observations and the need for an ability and willingness to interpret and act on those observations. The best thing I have ever read on that subject, and which I strongly recommend, is "Art and Science in Subsurface Engineering," by Ralph Peck, *Geotechnique*, March 1962.

8.3.2 Keeping Records and Writing Reports

One of the most onerous but very necessary aspects of field inspection work is keeping notes and records, some of which will be needed for periodic reports, written and oral, to the office and to the client. It may be that a tiny fraction of these notes will turn out to be of special importance when something goes wrong. The trouble is you don't know which fraction in

FIGURE 8.15 A rock coring box.

advance will prove to be significant. We could instruct an inspector to "write everything down" and be done with it. However, with the increasing tendency toward more and more lawsuits, such broad-brush, easy advice can lead to more serious problems than it may solve. So more careful consideration is necessary.

One way to view the problem is to separate it into three parts: how to keep fill records, where to take density tests, and what notes to write down or otherwise record (perhaps more importantly, what *not* to record). Let's take the easy parts first.

How to Keep Fill Records

The techniques that I adopted for the strip mine case study just described are, I believe, adaptable to many fill control jobs. First, obtain or prepare a large drawing of the site plan and profile (see Figure 8.6). Second, ask the contractor to install a cartesian system of poles on a 100-ft grid, as shown. (If you look closely at Figure 8.9, and others, you will see some of these poles.) Mark each of the poles with its location (e.g., c4). When density tests are done, it will be relatively simple to locate oneself on the site by pacing from the nearest pole. The determination of the approximate elevation of the test can be done using the Locke level as described in Section 8.2. With a little care, I would judge that one can be accurate to within a couple of feet in plan, and perhaps 0.5 ft in elevation, which is, I think, accurate enough for the purpose. The coordinates, elevation, and density result should be recorded in your pocket notebook, and subsequently plotted on the plan, with the decimal point of the density value serving also as the plotted point. The elevation can be recorded in parentheses, as shown by example in Figure 8.6a.

The technique described is helpful in that one can, at a glance, get a feeling for the general distribution of testing coverage, and adjust accordingly if some region of the fill is not being covered that should be.

A companion technique was the periodic running of a line of levels such as section 1–1 (Figure 8.6a) shown in profile in Figure 8.6b. On the job described, this was done (again using the Locke level) about once per week, usually on a Friday afternoon. The results were plotted, on successive weeks, on a profile, thus providing a simple graphic way of keeping the office and the client advised of the progress of the filling operation.

The job required weekly *and* daily written reports, which at the time I regarded as a little excessive, but inspectors don't make such decisions. I suspect that the unusual amount of reporting was caused by the equally unusual complications of the job. At any rate, copious and careful field notes were both essential and very helpful in preparing daily and weekly reports.

The circumstances of the job were such that the daily reports could not readily be typed. In such a case, the use of snap-off forms, with built-in carbons, can be very helpful and time saving. The retained copies are indispensable aids in preparing longer, formal reports at a later date.

As may be surmised from the photographs shown here, I strongly favor and recommend the use of an inexpensive camera for record keeping. Photographs provide not only a virtually objective and infallible record but may also save a lot of time and effort, especially when there is an unusual demand for written reports.

The use of a portable cassette recorder may also be a sensible option for record keeping and reporting. These electronic devices did not exist (at least at reasonable prices) when I did inspection work, but it occurs to me that their use might be feasible in some circumstances.

One last precaution: If the method of installing poles is used, it will be necessary to advise the contractor and the compaction operators to compact very close to the poles and to monitor this fairly closely. Without such insistence, the compactor operators are likely to miss the poles by a couple of feet on either side, resulting eventually in sizable areas of loose fill. If a footing happened to be built on such a spot, unacceptable settlement would likely result. One additional way of avoiding this event is to have the poles laid out with an origin such that no poles will be placed where footings are to be located. Unfortunately, footing location may not be known at the time of site preparation. Incidentally, the same logic holds if one is doing a test pit investigation of a site. Avoid digging test pits at a point where it is known a footing will be placed or is likely to be placed. For example, if the location of the corner of a proposed building is known, call for a test pit to be located at a point 10 ft outside the building, on an extension of one of the building lines.

On small jobs, the methods of record keeping need not be as elaborate as described, but on large jobs, a methodical approach is needed, supplemented by graphics and appropriate recording devices.

Where to Take Density Tests

The specifications may spell out the complete details of field density testing, including how many tests (per thousand yards, for example), and where to take the tests. Or the project engineer may instruct the inspector on how many and where. The other extreme would be almost no such instruction, in which case the inspector must depend solely on his or her own judgment.

If the project engineer has the responsibility and flexibility of deciding how to proceed, I think the decisions should depend largely on who will be doing the inspection work. If the inspector is judged to be experienced, intelligent, articulate, diligent, personable, but capable of forcefulness when necessary, I would stress more qualitative means of fill control, with only a minimum number of field-density tests. On the other hand, if the inspector possesses none of these qualities, have him dig holes all day.

My point is that one cannot do both at once. By qualitative fill control, I mean such activities as observing all aspects of the compaction process carefully, establishing methods, and conversing with and advising the contractor intelligently. If the inspector can do none of these well, then base the fill control on as many density test results as can reasonably be done in a day.

My preference, if I had the option, would be the former. Certainly, there would have to be some quantitative confirmation of the adequacy of the fill, in the form of some field-density test results, but I believe that the best overall job would result from taking advantage of the skills of the inspector by allowing him to spend a good deal of his time on the job exercising those skills, I'd say maybe only 25% of his day digging holes.

Where to take the tests? There are many ways to decide. A very simple technique, which has the dual advantage of explicitness and lack of bias, is described in Appendix F of the Asphalt Institute Manual (MS-1) *Thickness Design,* entitled "Procedure for Selecting Sampling Locations by Random Sampling Technique." The technique involves six flips of a coin. An example is given by Truitt (1983, p. 225). The example relates to selecting a horizontal location, but the technique could easily be modified to selecting random elevations. The technique has the added advantage of being very quick.

Of course, there will arise the necessity of taking tests that *are* biased, and this is where the judgment and other qualities of the field person are brought to bear, particularly the one of forcefulness when necessary. When, for various reasons (some predictable and some unforeseeable), the inspector has good reason to suspect a fill is deficient, the courage to insist upon testing at that location is essential. This may require excavation, and the contractor is not going to like it. I describe these and similar problems in more detail in Section 8.3.3.

Observational Notes and Records

The third category of record keeping is in many ways the most difficult. I have offered advice on some details of taking and recording load test data and field-density test data, and this is relatively easy because it deals with facts, usually of a numerical nature. But there will occur literally countless numbers of incidents, conversations, and discoveries in the course of a day that may not fit the description of "data" but that may be nonetheless important, and often much more important. I can only offer brief guidelines, some based on my own experiences, and some on the advice of others.

In a word, be as *factual* as possible.

Richards (1976) presents no less than 25 Rules of Conduct for the field engineer, and supports each with real, authentic case histories. I cite just two of them here, and add some commentary of my own. The first is "Cultivate Satisfactory Internal Relations" (Richards Rule 9):

The field engineer should avoid any *written* criticism either direct or implied concerning members of his staff or organization. Detrimental comments by the field engineer concerning his own organization may seriously embarrass him when produced in court.

The emphasis on the word *written* is mine. Also, I think Richards' advice is good without restricting it to "internal relations" or "his own organization." Certainly, it is worse to put criticism of your own firm in writing than of another, for you may wind up being fired on top of being embarrassed, if the document becomes detrimental in court. But I don't see where written criticism of anyone should be done unless you make the conscious judgment that it can *serve some useful purpose.* For example, if you find a discrepancy, or a "chink," or even an outright blunder in a written specification, *phone* your project engineer first and raise the issue diplomatically. As an example of an external miscreant, if a contractor is regularly exceeding the specified fill thickness, *don't* write "every time I turn my back, he doubles the allowable lift thickness." Instead, write "three consecutive lift thicknesses were measured as 18 in., 20 in., and 22 in. The contractor was advised that this violated the specifications."

The latter constitutes a combination of my advice on being factual, and Richards' Rule 9.

As a hypothetical example of reporting a dilemma overlapping professional, ethical, and moral obligations, consider the following: You are a professional engineer of some experience. You pass a construction area (let's say in the New York metropolitan area) where you observe some men working in a 10-ft deep, unsupported trench in soil. What do you do?

This is a good one to kick around over coffee. I have done so with a colleague and we concur that one should (1) ascertain quickly who the controlling agency is, (2) call them immediately and alert them to the safety violation (OSHA) you observed, and (3) follow up with a factual letter (*not* a sanctimonious diatribe or lecture).

While the foregoing is hypothetical, one can see that it could very easily happen. I shall never forget attending a Friday lecture in soil mechanics at Oklahoma State University on stability of vertical, unsupported soil excavations. Over that weekend, a worker was buried alive and suffocated on a sewer job on the edge of town.

And so one can see that the field person is constantly faced with a wide variety of decisions regarding what to report, whether to do so orally or in writing, and what words to use, particularly if the choice is a written report or notebook entry. It is not easy.

Richards' Rule 25 is:

Maintain Adequate Records: The primary task of the field engineer is to be able to reproduce at a later date project events. Adequate records of project construction activity, conversations, visitors, telephone messages and other related project events will be maintained by the reactive field engineer.

That's quite a mouthful. Like Richards, I can't define exhaustively or precisely what "adequate records" are, or what "other related project events" might be, but the above advice combined with the caveat to be as factual as possible, will hopefully be of some help.

Finally, when in doubt, talk to your project engineer or other more experienced person whose judgment you respect before putting something in writing. But if you stick to facts, you'll rarely be in doubt.

8.3.3   Contractor–Inspector Relationships

Another sticky problem one has to deal with is that of contractor–inspector relationships. In the previous section, I described the ideal field person, using the words personable and forceful (when necessary). The reason such characteristics are important is that the field person has to be able to deal with all kinds of contractors equally as well, in terms of ensuring that the specifications are followed. Conversely, the contractor has to deal with all kinds of inspectors and other problems so that, as Morris has written, he "can complete his work with pride and still make a reasonable profit."

Before proceeding further, so that no one will get the impression that I an unduly biased, let me say that it is my opinion—and I believe that it is the only one that makes common sense—that there are in this imperfect world of ours "good" and "bad" people in every profession and walk of life: engineers, lawyers, clients, contractors, and inspectors. And we must try to deal effectively with each other. Because of the fact that my experience, as it relates to writing this section, has been as an inspector dealing with contractors, it follows that most of what I have to say is about these relationships. And despite the fact I have not had extensive experience, I believe I've been lucky (if that's the word) to have had experiences with a wide variety of people: the good, the bad, and those in between. In addition, I take a lot of notes! But you be the judge.

The reason I chose "personable" and "forceful" as pertinent adjectives to describe the good inspector is that these traits are needed to deal effectively with the "good" and "bad" contractor, respectively.

An example of a good contractor was the one I dealt with on the strip mine job that I described rather fully earlier. (Actually, he was the earthwork contractor's superintendent. Throughout this section, I shall use "contractor" to mean the person in charge at the site with whom I dealt directly as an inspector.) He was good in the sense of Morris's description in that he had pride and a natural wish to perform well and make a reasonable profit. This was very fortunate for me, and we worked together very well, had a thoroughly amicable relationship, and got the job done effectively in spite of a most unusual set of problems.

One of the contractor's deficiencies, however, was a surprising ignorance of even the rudiments of soil compaction technology, notably field testing and how it relates to the whole fill control operation. I found this to be a rather common deficiency among contractors. So, unless my experience was somehow not representative, and unless things have changed drastically, an inspector should not be surprised by this finding.

This lays the groundwork (no pun intended) for my first suggestion for dealing effectively with such a contractor. *Make a conscious effort to find specific, legitimate ways of helping him perform well and increase his profit.* I can think of three ways this was accomplished. First, when the test strip density testing was done, I showed the results to the contractor, and we agreed together to use the interpolated result of six passes. An alternate approach would have been for me to simply state "we'll need seven passes." While I never discussed this with the contractor, I truly believe that he thought "gee, he's not out to do a job on me," or words to that effect. Subsequently, on one of the few rainy days on the job, I suggested that we could keep working effectively if we immediately compacted the soil upon placement, that even a five-minute delay in the fairly heavy, steady rain would wet the loose lifts to a degree as to make compaction impossible. He agreed, and we converted another Euclid truck to compaction duty and were thus able to work a couple of more hours. The third suggestion was to use rock excavated from the rear (cut) area of the site as fill, breaking the rock with passes of heavily loaded tandem sheepsfoot rollers, followed by a lift of soil borrow to fill the void spaces in the rock. The initial thought was to waste the rock fill.

Each of these suggestions was well received by the contractor, for obvious reasons, and each was regarded as compatible with the specifications and toward the effective and economic completion of the job. Taken individually, none of the suggestions was a big deal, but collectively they served the purpose of harmony and maybe even a little pride all around. With all of the unusual problems of this job, I hate to think of what it would have been like if the contractor had been "a bummer." I believe that if you perceive the contractor to have the characteristics described, the approach and attitude of consciously seeking ways of helping the contractor with sound advice will pay large dividends. I don't expect he'll be aware of this strategy unless he also reads this book from cover to cover, and I don't expect that every earthwork contractor and superintendent is going to rush out to buy this book—though that would be nice. I am not suggesting here that the inspector curry favor with the contractor, but rather that he foster harmony by offering legitimate, sound, money-saving advice.

And now for the other side of the coin. I was assigned briefly to a fill job in northern New Jersey. While I wasn't familiar with the details of its administration, I learned from conversations that it was a rather splintered job, involving a general contractor and an unusually large number of subcontractors, one of which did the earthwork. Coordination and scheduling on such jobs is difficult under the best circumstances, but on this job the circumstances were a nightmare. Early on, the job was plagued by criminal vandalism, including sand poured into gas tanks, and vertical number 9 reinforcing bars, encased in footings, smashed into U shapes. As a result of the early problems, the schedule of construction operations got completely out of control.

As described earlier, the general contractor requested an unusual procedure to attempt to improve or restore scheduling; he asked for approval to raise the perimeter of the main structure to bottom of footing grade, to allow construction of the exterior footings. Although this was not good practice, since it would create, in effect, a diked interior, whoever was in authority approved the request, I suspect in sympathy with the general contractor, a good man in a bad situation. The idea was that the interior would be filled later. (With the general contractor's luck, I half expected a major rainstorm before this could be done, making a king-size lagoon of the site.)

Because only a narrow perimeter strip would be filled, the elevation of the fill could be raised rather rapidly. To further speed things along, the earthwork contractor got approval to bring in a lot of extra equipment, mostly scrapers, on a Saturday. I shall never forget that Saturday.

Earlier in the week, when filling of the perimeter was being done at only a moderately accelerated pace, I was already having difficulties with the earthwork contractor, principally with excessive lift thicknesses. I spoke to the contractor about it repeatedly, and soon perceived that he was "trouble." You'd get a "Yeah, OK," but as soon as you would turn your back, he'd signal to the scraper operators to "lay it down." I often wished that I had

had a long-lens camera to catch him with his hands separated from his knees to a point above his head, in the posture of signaling the scraper operators. I had called the office about the problems, having recognized that I was dealing with an incorrigible. In such cases, a call from the office or a visit to the site by the project engineer can have some positive effect. On Saturday, the situation got almost out of hand. As noted, extra scrapers were brought in, but no commensurate extra compacting equipment. Scrapers were whizzing by, and more than once I was engulfed in dust clouds, and was actually a little nervous about being run over. (Scrapers have been known to tip over. See Section 8.3.4, following.) As you might expect, the contractor was busy with lift thicknesses, and the day was filled with arguments and recriminations. This was to be my last day on that job, as I was reassigned to another.

In a situation comparable to the one just described, an inspector has to muster the courage and forcefulness to assert himself. But do it in stages. First try to handle it yourself, rather than calling the office at the first sign of trouble or an uncooperative contractor. Keep written, *factual* records of conversations, violations, and assertions of violations transmitted orally to the contractor. At some point, inform your project engineer and seek advice and, possibly, assistance. One of the approaches I have heard about is to threaten to walk off the job, and then do it. This would be done only as a last resort, and when other circumstances are compatible with the potential efficacy of such action. For example, if the specifications or some regulating agency require certification before payment, you can make the point that the fill cannot possibly be certified if you leave the site. Such pressure can work wonders.

Assertiveness and forcefulness may also be necessary in inspecting fills in confined areas such as trenches, and behind basement walls, where large thicknesses of fill can be placed (or dumped) in a short period of time. The specifics of these problems were discussed more fully in Section 6.5.2, but it bears repeating here.

Interpersonal Relations

It is important that the inspector not become overly friendly with the contractor on a personal level, to the point of socializing. I made this mistake once. On a drilled pier job, where it was my responsibility to inspect the rock surface at the bottom of the belled pier for proper size and rock quality, I got on friendly terms with the drilling contractor. He invited me to a cookout at his home on the weekend, and I accepted. This was a job where commuting was not feasible, and I stayed in a hotel in the area, so a weekend cookout was an attractive invitation. A couple of days later, I was compelled to reject a pier bottom after the drilling rig had already been moved to another location on the site. This rejection was prompted by the fact that the rock was of the unusual type that would deteriorate very rapidly when exposed, as I discovered, even overnight. What happened was that I had inspected and approved the pier bottom in question late one afternoon, but the concrete delivery service had quit for the day. Upon reinspection the next day, the rock was found to have deteriorated to the point where you could pull out chunks, with some effort, with your hands.

I remember two things from this incident: my reluctance to tell the contractor about the need to redrill and the look on his face when I did. (To maneuver the rig back onto the hole is often problematic and always time consuming, so the contractors are not happy to do so.)

What I learned from this is not to get too friendly with a contractor, because even if you do so innocently and with no intent to have it affect job decisions, there will be an almost unavoidable influence when it becomes necessary to reject some work requiring corrective action. Similarly, the look on the contractor's face suggested to me that he thought "I just fed you a huge steak, and you do this to me?"

Richards describes some authentic case histories regarding matters such as accepting direct favors and gratuities, ranging from bottles to ball games to business lunches. His Rule 2 is, "Constantly be discreet. Never accept favors or gratuities." He also describes cases of outright bribes, some of major magnitude and national notoriety.

The opposite extreme of the friendly inspector is one who is rigid, cold, and totally impersonal in dealings with the contractor. This type can be regarded by the contractor as a nitpicking pain. Such an extreme stance is, in my view, not a good idea either, for it can impede the development of harmony such as described on the strip mine job. And sometimes the cost of harmony is to give a little. One last personal experience will illustrate what I mean. I was assigned to a pile driving job, and the specifications called for a cushion block for the butt end of the pile to be a block of wood cut from a tree trunk and trimmed to fit. (This is to prevent damage by the direct impact of the hammer.) The contractor instead used split chunks of wood, as I recall, each about as big as your head; before the start of driving, a worker would pitch a number of these into the cylindrical space above the pile butt. For several days, I observed this procedure and raised no objection to the procedure, despite the fact that I was well aware that it was, technically, a violation of the specifications. In due course, one of my bosses saw or heard about this procedure, and became very upset because of the violation and my allowing them to use "chips" for the cushion block. Not being bashful, and perhaps a little imprudently, I argued with him privately that there was, in my view, no practical significance to this variation in procedure, that after two or three initial impacts of the hammer, there would be no practical, distinguishable difference between the cushion block material, chunks vis-à-vis a one-piece, trimmed block. I pointed out that I would not countenance the practice of throwing blocks into the rig near the end of the driving of the pile (called fetch-up), as this could seriously diminish the actual carrying capacity of the pile. (A contractor doing the latter would be guilty of intentional cheating, the purpose being to reduce possible damage to the pile hammer by improperly cushioning the pile during fetch-up.)

What we had here was a difference of opinion on two issues: the enforcement of specifications and whether or not the one-piece block or the use of chunks, as described, made any significant difference. I believe the latter could never be determined in any scientific way, and thus will have to remain a matter of opinion. However, the broader and more important issue of specification enforcement is worth considering. It is no doubt the safer and more conservative approach to adopt a "specs are specs" rigid approach, since no decisions would ever have to be made. I admit that it may be controversial, and in some circles almost heretical, but I believe that one always has to make adjustments: not, "Are the specs being followed?" but "What is best to produce the best job?" No specification ever written, or that will ever be written, is perfect. Indeed, as indicated by the case studies in Chapter 4 (especially Case Study 2), many are very badly written or pieced together by people with little knowledge of the technology, necessitating major changes, legally or illegally. If a minor variation is permitted, and the result is better harmony on the job, what's the harm? The alternative can be hostility, day in and day out, between the inspector and the contractor, and I can't help but believe that this will somehow, but with certainty, result in a poorer job.

In retrospect, as an inspector, what I should have done was call the office about the chunk issue when I first observed what the contractor was doing, offered my opinion, and then accepted their decision, irrespective of whether I agreed or not. But frankly, at the time, I didn't think it was an important enough issue to bother the office. I don't remember the name of the boss, and he doesn't know this, but I have referred to him as "Chips" lo these many years. Who says field work isn't fun?!

8.3.4   Odds and Ends

Two items of some importance that did not seem to fit neatly into the chronological description of the strip mine job are included to close this chapter. One deals with personal safety, and the other with rock excavation.

A Note on Personal Safety

Figures 8.8 and 8.16 show scrapers placing soil and picking up soil, respectively. Notice in Figure 8.8 that the scraper has a high center of gravity, as evidenced by the concentration of mass in the area immediately above and behind the driver. Now examine Figure 8.16, and imagine the scraper coming down the slope at about 35 mph or better. The combination of three factors can contribute to a significant chance of rollover: the high center of gravity; steep, erratic slopes (perhaps with partially buried boulders); and the "cowboy" attitude of many scraper operators. The latter explains the caption of Figure 8.16.

FIGURE 8.16   A scraper rodeo.

I spoke with a scraper operator about this on one job and learned that, indeed, they are known to go over occasionally. He told me of one such recent rollover that he knew of where the operator sustained a serious head injury. Some other interesting tidbits: Some scrapers are made with cabs that serve as roll bars, preferred by the older, heavier, or otherwise less agile operators. The younger, agile types choose the cab-free type, the better to jump clear if the occasion demands. I was also told that there are no *old* scraper operators, because the kidneys and other inner organs can't take the abuse after about age 35, presumably assuming 15 or so years of prior experience operating a scraper. In Figure 8.16, note also the bulldozer assisting the scraper with the excavation in the dense soil.

A Note on Rock Excavation

As the cut proceeded at the rear of the site of the strip mine job, rock was encountered at shallow depths beneath the glacial till, necessitating both ripping and blasting. The shallower rock yielded to the ripping operation shown in Figure 8.17. This was the rock that was placed in the fill with alternating layers of soil fill, as described earlier. As we got deeper in the rock, the rock quality increased (see Figure 8.15), and blasting became necessary. Fortunately for purposes of cost containment, only one blast was required. See Figure 8.18. (Since I took this photo from the next county,

FIGURE 8.17 Rock excavation by ripping.

FIGURE 8.18 Rock blasting.

I think you will agree that my timing is to be admired more than my courage.) Apparently this loosened the rock sufficiently to resume ripping effectively. These photos and the accompanying comments are included to futher emphasize and implant the importance of rock excavation as a construction cost.

8.3.5   Quiz Answer

Did you peek?

Answer: Yes. There would be no differential settlements only if all footings were of the same size, which would require that all column loads were the same, a virtually impossible case. See Section 8.1.1 for further description.

8.4   Glossary

Tectonic (forces).   Internal geologic processes that influence surface events. See any modern book on physical geology for an update on the new and emerging discoveries in this fascinating field.

Varved clays.   A series of repeating thin soil strata, each series representing a one-year cycle of alluvial (lake or *lacustrine*) deposition. Clay laminae (thin layers) represent winter deposition, and subsequent sand–silt layers represent the following spring (runoff) deposition.

Tare.   The container in which a material is weighed; hence tare weight.

9

Techonomics

Having discovered the fun of inventing new words early in this work (gem-icoss), I thought I'd try again with the title of this short chapter. It is short because it is a subject about which I know very little, largely because I have spent most of my professional career in the academic world where one does not deal very much with the business and economic aspects of consulting, construction, and interaction with clients regarding business economics. I will also say that I have always had a sensitivity toward, and appreciation for, the important link between technology and economics.

What I have to say comes from this interest, but largely supported and augmented by conversations that I have had on these subjects with professional colleagues.

9.1   Engineering Design

The conversation I had that prompted me to think about adding this chapter was with an executive of Woodward–Clyde Consultants (geotechnical specialists). By interesting coincidence, this firm is the same one—now grown considerably larger—where I had my start in professional geotechnical engineering, and where almost all of my early experience and interest in compacted fills was obtained. So it is, I think, quite fitting that this chapter came about in that way, sort of a closing of the cycle. A concern was expressed that too many engineers were not sufficiently sensitive to how economics *should* influence technical decisions. A specific case history was cited as an illustration.

A foundation investigation was made for a warehouse for a large greeting card company. The warehouse site was in a locale where the subsurface conditions were very poor, with soft compresssible soils to substantial depth.

After appropriate sampling, testing, and analyses, the engineers recommended the site stabilization technique known as "surcharging with sand drains," a now-common procedure where vertical sand columns (or, more recently, artificial "wicks") are installed in a designed grid pattern to appropriate depths, and a surcharge load is placed on the site to induce drainage (preconsolidation). The load is created by a fill of specific design thickness. The process results in settlement and increased strength of the subsurface soil. The typical design time for such stabilization to occur is about one year. The surcharge is then removed (usually) and the structure is then built on the preconsolidated, stabilized site.

As was pointed out, however, while the design was technologically correct, it was done without any economic considerations. First, the "cost of money," as it was phrased, was about 15%, and second, the one-year period would deprive the client of the use of the warehouse for a full December holiday season, a period when this particular client's profits are obviously concentrated.

Until and if interest rates come down significantly, it would appear that the surcharging procedure might not be viable for any projects, certainly for a lot fewer than in the old days of 6% interest rates. The second consideration is, of course, unique to this client. Such considerations dictated a pile foundation, in that the interest cost would be largely eliminated and the warehouse would be completed for use by the holiday period. Since I had this conversation, it has occurred to me that the use of lightweight fill in a weight credit application might have proved to be a viable alternative consideration, as was described in Section 5.3.1.

9.2   Legal Costs

Another aspect of techonomics that deserves some additional comment is legal costs. I have discussed some of the issues with colleagues and done additional research to further educate myself for the preparation of what follows. Actually, long before I contracted to write this book, I was interested in how legal matters affected my personal economics in terms of part-time consulting activity. So my studies precede research for this book.

The nation's media deal increasingly with the burgeoning problem of the growing number of lawsuits and their cost. Most recently, Herb Jaffe (*The Star Ledger,* Newark, NJ, January 22, 1985) referred to "a steadily rising lawyer population . . . new laws [generating] lawsuits . . ." contributing to "the litigation nightmare of the 1980's." One statistic that suggests the proportions of the problem is that more than 40,000 new lawyers a year graduate from 175 accredited law schools in the United States. Increasingly larger judgments, lawyer's fees, expert witness fees, and court costs, combined with the much larger number of lawsuits, drives the cost of professional liability insurance to prohibitive levels.

Attorney's fees and expert witness fees are often very high. In a recent newspaper description of the settlement of the Agent Orange case, one lawyer was reported to have requested the eye-opening rate of $450 *per hour*. Expert witness fees, while usually not in that range, can also be impressive. Such fees are typically substantially higher than hourly rates charged for standard engineering design services. I have often thought it anomalous that this is so. It has always seemed to me that it is easier to figure out what went wrong afterward than to avoid error in design or construction. On top of that, there's less risk! While I suppose it happens, how often do you hear about an expert witness getting sued? To make it more anomalous, the witness doesn't even have to be right, he just has to make the most convincing case.

I recall an interesting and fascinating case in point. Some years ago, G. A. Leonards, of Purdue University, delivered the annual Terzaghi Lecture, which dealt with a major pile foundation failure (in Africa, as I recall), for which he was the eighth of a succession of experts. I vividly recall Professor Leonards' last slide, a summary of the eight explanations of the cause of the failure, all of which were (I recall) mutually exclusive.

Thus, a case can be made that the costliest engineering work is easier, less risky, and you don't even have to be right. (In fairness, perhaps it should be added that the *pressures* of such work can be substantial—like eight hours of hostile questioning.)

As other observers have noted, legal costs are passed on to clients. It is also likely that direct engineering costs are also higher because of an understandable tendency to do more rather than less sampling and testing because of the fear of lawsuits, rather than for strictly technological reasons.

9.2.1   A Proposal

I have a specific suggestion for at least a partial solution: a concerted effort directed towards the elimination or substantial reduction of the attorney's contingency fee system in the United States.

What prompted this idea was an interesting and intriguing discovery that this system is not used in Canada. I learned this from a document that was loaned to me by a colleague with concerns on the subject similar to mine, Professor Ed Dauenheimer of the New Jersey Institute of Technology. Actually, his concerns are naturally greater than mine, since he is a specialist in construction planning, having held professional engineer's licenses in 18 states, with extensive experience in engineering construction practice, and presently engaged in part-time consulting. The document is a pamphlet entitled, "Professional Liability Loss Control: Architects and Engineers." It is printed and distributed by INAX Underwriters Agency, Inc., Chicago, Illinois, a company that advises clients on how to minimize their exposure and losses from lawsuits.

In discussions with Professor Dauenheimer I learned that concerns about the costs of potential lawsuits have become so pervasive that (1) liability insurors actually dictate to potential clients what language, and even words, may and may not be used in contract documents, or else they will not provide coverage (maybe that's how "observers" originated), and (2) INAX provides clients with a thick looseleaf binder, mostly composed of *Do's* and *Don'ts;* according to Professor Dauenheimer, "mostly Do Not's."

In my experience as an expert, the opposing lawyer always asks, "How are you compensated for your services?" and I reply, "On an hourly fee basis." The implication is clear: If you're paid on a contingency basis, your testimony will be portrayed as biased because of your financial stake in the outcome. I have never understood why the same logic does not apply to attorneys.

The contingency fee system, as stated in the INAX pamphlet, allows a plaintiff to sue with little or no expense. Thus, the plaintiff has everything to gain and nothing to lose by suing, and it is becoming more and more common for lawyers to advertise their services using the contingency feature as the principal focus of the advertisement. The INAX pamphlet further states that "conversely, in Canada, a plaintiff must assume a substantial expense in filing a suit, since the attorney is paid a flat fee or retainer. . . ."

What specific steps can be taken to eliminate or modify the system? First, I'd suggest that a coalition of some sort is needed, one with much clout, since lawyer's groups would doubtlessly provide formidable resistance. I hope, however, that many lawyers and judges would agree with at least some of the arguments presented here.

The coalition could include professionals from The American Society of Civil Engineers, The National Society of Professional Engineers, The American Institute of Architects, and contractors' associations. Because of the broad national scope of the problem it might be feasible, and it certainly would introduce more clout, to attempt to include the Ralph Nader organization.

Second, find out how Canada does it. Since Canada is a free society like ours, I presume it's not through government control. So it is likely that a self-regulatory body is responsible. If that is the case, that information could be used as pressure to reform the American system.

Perhaps some intermediate solution is possible and desirable, wherein a system could be developed by the American Bar Association to impose limits of some sensible dimension on its members, much like the ban on advertising (recently relaxed). This might be desirable because of the valid argument that people of limited means would have no legal recourse without a contingency system. Perhaps state boards could be established to consider applications from aggrieved parties who could not otherwise afford to sue. Again, what does Canada do in this regard?

It would be interesting to compare the claims statistics of the United

States and Canada for the same time period, assuming the data for the latter could be obtained. (The U.S. statistics for most of the decade of the 1970s are presented in chart form in the INAX pamphlet.) Perhaps that would provide some indication of the effects of the contingency system. A valid and meaningful study of such a complex problem would need to be much more comprehensive, since many variables other than contingency fees are involved, but this would be a start. This would be a great topic for a social science major.

10
Update at Press Time

This final chapter includes descriptions of material discovered since the completion of the manuscript. Some of it is still literally in the process of evolving. The most significant new information has been in the area of lighweight fills. This permits still another alternate approach to foundation design problems in weak soils by using the principle of weight credit. (See Section 5.3.1.)

Also included are sections on very large jobs, and one entitled Annotated References for Self-Study. There appears a list of 49 standard references that were used (and cited within the text) for the preparation of the book. It has occurred to me, however, that some additional references, not cited in the book, would be helpful to readers wishing further information on specific subject matter not fully covered here. Field load testing, for example, is only mentioned peripherally in the text, since a full treatment would not be consistent with the book's focus. Nonetheless, since one corollary purpose of the book is to provide what amounts to a field manual (notably Chapter 8), the same young persons interested in the field work described would also be interested in all other aspects of field work. The self-study references (including some excellent sound film strips) are listed here to accommodate these young persons, and their busy supervisors.

Finally, to augment the suggested self-study materials, I suggest my own list of seminar topics. Many topics cannot be presented adequately, solely in written form, and are most effectively presented orally, with audiovisual aids.

10.1 Weight-Credit Update

Figure 10.1 illustrates the combined use of two super lightweight fills, pre-cast rigid boards and a material called Poleset. Both materials are foam

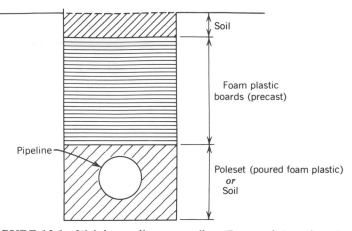

FIGURE 10.1   Weight-credit versus piles. (Rospond Associates.)

plastics, but of significantly different composition. From my perspective as a foundation engineer, a principal difference is the method of placement. Poleset is a poured-in-place material. One of the important distinctions, as mentioned briefly in Section 5.3.1, is that Poleset would seem to be much more suitable for filling irregular and/or confined spaces. That was the initial rationale used in stipulating the poured Poleset backfill under, over, and around the pipeline of Figure 10.1, with the precast boards simply stacked in place, above.

10.1.1   Plastic Boards

Styrofoam HI, one of the available plastic boards, is a registered trademark of the Dow Chemical Company. Some of its important properties, including cost, have been described in Section 5.3.1. Its use in weight-credit foundation construction is also described in Chapter 5. It is covered by two patents (references 29 and 30). Figure 10.2, the Pickford Bridge, shows the stacked bundles of Styrofoam HI boards. Details of the job are described in Section 5.3.1.

More information on an assortment of rigid foam plastics, their properties, and current costs and availability may be obtained from Research Specialists, Dow Chemical U.S.A., P.O. Box 515, Granville, Ohio 43023.

10.1.2   Poleset

The information I have about Poleset is sketchy, obtained initially from nothing more than a blurb of about 200 words in an article in *Better Homes and Gardens*, April 1974, p. 60. Most of what I have learned is very promising: It is very strong, very lightweight (quoted at $1\frac{1}{2}$ lb/ft^3), and is reported to be waterproof. Variations in mixing ingredients can produce a wide va-

FIGURE 10.2   Styrofoam HI approach fill, Pickford, Michigan.

riety of strengths, densities, and costs. This is consistent with what I had learned about foamed plastics earlier, especially the quotation from *The Modern Plastics Encyclopaedia (1968):*

. . . foams may be produced which have densities ranging from less than one pcf to about 70 pcf, with an almost limitless range of chemical and mechanical properties.

Poleset is supplied by Forward Enterprises, Inc., 9430 Telephone Road, Houston, Texas 77075.

The job depicted in Figure 10.1 is for a site in Woodbridge, New Jersey, underlain by very soft, weak, compressible soils. Soils consultants had recommended piling for all structures, *including the utilities.* (The latter is italicized to emphasize that this is not very common practice and would only be needed for exceptionally poor soil conditions.) Having been earlier made aware of my weight-credit methods, the project engineers conferred with me, and the investigation suggested by Figure 10.1 began. This led to a computer study by Rospond Associates, which compared the costs of weight credit versus piles. Weight-credit, I am informed, was cheaper by $170,000.

As it developed, Poleset was not used in Rospond's design because the boards provided enough weight credit so that normal soil fill could be used around the pipe. I do not have any cost figures yet for the Poleset, but according to Rospond, "it's expensive." With little doubt then, it is currently more expensive than the plastic boards. However, the laws of economics (higher volume, lower unit price) might tend to make Poleset more competitive in the future, particularly if its (apparently) promising mechanical and physical properties make it especially suitable for increased use as a lightweight fill in weight-credit applications.

As a result of a presentation I gave to the Geotechnical Group of the Metropolitan Section, ASCE, an application of Poleset is being considered for a job in Jersey City, New Jersey. The floor slab of a very old structure is virtually unusable for significant floor loadings due to the excessive settlement of an underlying loose miscellaneous fill. The tentative plan is to remove about 4 ft of the fill, judged to weigh about 90 pcf, and backfill the space with poured Poleset.

Increased interest in lightweight fills for weight-credit applications was also generated because of considerations of the multibillion dollar Westway project (a planned interstate highway section on the west side of Manhattan that would have incorporated many satellite projects in land and real estate development.) This partially explains the huge cost of the federally financed project. Since the new land, about 225 acres, would have been created by filling along the eastern shore of the Hudson River (as with the earlier 93-acre Battery Park described in Section 5.3.1), the usual dredging and filling process would involve placing river muds on river muds. Because of the geologic processes involved in their formation, such muds are almost always fine grained, soft, relatively heavy, and thus potentially very compressible. The fill, of course, would be carefully compacted under the supervision of geotechnical experts, but the soft, natural soils below would undoubtedly create concerns about settlement. The selected use of weight credit, for certain pieces of the project (i.e., not major structures), might often be an alternative design possibility, an alternative to piles, for example. In fact, it was nine years ago that I conferred with Robert Johnston, of Mueser Rutledge, Wentworth, and Johnston, New York Geotechnical Consultants, about the possibilities of using Dow's Styrofoam for just such a piece of Westway. It was he (quoted in Section 5.3.1) who said, then, "I hate to bury voids."

At press time, the status of Westway, or a substitute project, remains in some doubt.

10.2   Very Large Jobs

With the exception of very brief descriptions (subtitled "Specification Components") in Section 7.1.1., the problems that are characteristic of very large compacted fill jobs have not been included in this book, for three reasons. First, I have endeavored to focus on describing first-hand experiences, and mine have been limited to small jobs. (As will be explained, "small," and "large," and "very large" are very subjective descriptive terms.) Second, the book is directed mostly to nonspecialists and aspiring young and inexperienced geotechnicians. Finally, it is inconceivable that jobs of the magnitude I am about to describe would (or could) be considered without the careful and expert input of geotechnical specialists.

Because of these reasons, I will provide only a summary description of the project highlights. Fortunately, an excellent paper has been published, and was recently brought to my attention by its senior author, Dennis J. Leary, P.E., who served as the project engineer. The paper is entitled "Earthwork Quality Control for Ludington Pumped Storage Reservoir," published by the University of Wisconsin—Milwaukee and the American Water Resources Association. The paper was presented at the International Conference on Pumped Storage Development and Its Environmental Effects, Milwaukee, Wisconsin, 1971.

The project involved the construction of a six-mile-long earth embankment (the upper reservoir) on the East shore of Lake Michigan, which included five major embankment soils (textures) from a common borrow. The total volume of the embankment was 39 million cubic yards, placed at rates as high as 180,000 yards per day. (Contrast this to the job, Case Study 1, Section 4.1.1, where the *total* fill volume was 150,000 yards.) In addition, a compacted clay impervious lining, 5–8 ft thick, was constructed. Filter sands were hauled approximately 40 miles by truck and base drain gravel was imported from the Upper Peninsula at the rate of three ships per week.

Two teams were involved with the work: a quality control inspection staff composed of a resident engineer, two field engineers, and 26 inspectors; and a quality control field and laboratory testing staff composed of the soil engineer, 4 engineers, and 14 field and laboratory technicians. The latter team was divided, interchangeably and flexibly, into three groups: field testing, laboratory testing, and data processing.

Methods specifications were used, including the use of more than 50 vibratory sheepsfoot compactors, rubber-tired tractors, and static sheepsfoot rollers (for the clay lining). Loose lift thicknesses, coverages (passes), maximum roller speeds, and water contents were stipulated for each soil type.

Field tests included in-place density tests (Washington Dens-O-Meter), loose lift thickness measurements, and static cone penetrometer soundings. The latter tests were done to evaluate and thus ensure uniformity of compaction, augmenting the field-density test results.

Visual and manual soil descriptions and classifications were made by field technicians and verified by laboratory tests. Index property tests included liquid and plastic limits and grain-size distribution determinations. Engineering tests included the standard Proctor compaction test and the relative density test (depending on fines content), and permeability tests of the various embankment zone materials.

As may be inferred by the magnitude and scope of this work, the principal distinctive feature of such jobs may be summed up in one word: management. Management of the personnel and logistics that are involved, and management of the enormous amounts of data that are generated. Implicit in management of personnel and logistics is the setting up of organizational charts, modes of communication (especially systematic reporting), the training of individuals, and the organization of individuals into working teams. The analysis of test results, especially where voluminous data sets are involved, is increasingly being done by statistical methods, and this was done for the Ludington project.

Methods of determining and ensuring representivity of test results were done in an interesting and perhaps unique way: "Density tests were required to meet criteria concerning depth, distance from unconfined slopes, and soil conditions."

As stated by the authors, "tests made too near the surface give low and unrepresentative results." This observation corresponds with my commentary about density gradients and their investigation (Section 5.5), and the need to "prepare a fresh surface of the compacted fill . . . by scraping off the top few inches of the rolled fill to eliminate effects of shrinkage by desiccation [and other surface disturbance]" (Section 8.1.2). I was interested to learn that these effects were taken into consideration in the Ludington project. More interesting is the fact that a method was developed (but not explained) for choosing a specific test depth for different soil types (varying from 0.5–2.0 ft).

The requirement for "minimum distance from unconfined slopes" (for the location of field-density test holes) is, to my knowledge, unique. It is a very logical requirement. (In fact, when I read the paper, I thought "why didn't I think of that?") For the Ludington project, the specified minimum was 5 ft from unconfined slopes, and 10 ft from "unconfined corners" (e.g., in sand blanket near the intersection of longitudinal and lateral [orthogonal] drains). The authors state that "these distances were established on the basis of test results obtained at the start of embankment construction." Should readers wish to obtain further information regarding these criteria, or other aspects of the Ludington project not covered in the published paper, contact Dennis J. Leary, P.E., Langan Engineering Associates, Inc., River Drive Center 2, Elmwood Park, New Jersey 07407.

The third criterion, soil conditions, is directed to the not-uncommon occurrence where the texture of the soil removed from the density test hole differs significantly from that of the laboratory sample that was used for the compaction test. The authors state (in such a case), "the hole is abandoned." This is in accordance with commentary I have made with respect to simulating field conditions (Section 8.1.2). Unfortunately, neither Leary nor Monahan have established any quantitative criteria for determining, with any meaningful exactness, *when* to "abandon the hole."

10.3   *Annotated References for Self-Study*

To provide additional help to young persons starting in any of the various aspects of engineered construction, I recommend the following for self-study. The acquisition of these books and audiovisual materials by employers will also serve to minimize the time and expense for supervisory and training activity. One caution to these people, however: Review the materials. While everything I list here is generally recommended, I do have some reservations about some of the advice given. There are some things in some Soiltest filmstrips with which I disagree, for example. A suggested approach is to view the filmstrips initially with the staff people, review the work afterward, and perhaps prepare a written critique detailing differences, additions, or corrections. Such a document could then be distributed to future viewers (new staff) as an addendum to the filmstrip.

10.3.1   Soiltest Filmstrips

As a part of the laboratory component of the undergraduate course in soil mechanics that I developed and taught over 26 years at Newark College of Engineering (NCE), these Soiltest sound filmstrips were required viewing during periodic scheduled laboratory write-up sessions:

1. Auger borings (including hollow stem).
2. Split-barrel sampling/standard penetration test.
3. Thin-walled tube sampling (shelby tube).
4. Field-density testing—sand cone method.

These were scheduled for viewing by student laboratory groups during periodic write-up periods common to undergraduate laboratory courses. Each filmstrip runs about 20 to 25 minutes, so the entire three-hour write-up period was not required. The filmstrips were chosen to allow some coverage of important field operations necessary to geotechnical work. Prior to this solution, I wrestled with the problem of how to get this type of information across to the students, since I judged that it would be inappropriate to spend significant amounts of time on such subject matter in classroom lectures, where soil mechanics theory was, appropriately, of highest priority. (At NCE, students take the lecture and laboratory components of the course in the same semester, three hours each per week, for 15 weeks.)

Geotechnical specialists will recognize the importance of students gaining at least some knowledge about the operations of the filmstrips listed, but it would be beyond the scope of this book for me to include anything but the most general observations about the content, remembering also that the focus of this chapter is self-study.

Two additional filmstrips I showed in the laboratory course were "Atterberg Limits Testing" and "Unconfined Compression Testing." These were shown in the early part of the course, after orientation sessions and some hands-on experience in textural classification (grain-size distribution by visual and tactile inspection, by sieving, and subjective evaluation of plasticity of cohesive soils, all by Burmister's field inspection techniques). For efficiency and convenience, these two filmstrips were scheduled for viewing by the entire class on the day scheduled for the respective tests. As I recommend to supervisors, I would follow up the filmstrip viewing by about a 15–20 minute commentary on what the students had just seen, thus modifying the content of the filmstrips to my own purposes. The students scheduled for the test that day would then move to the adjacent laboratory to perform the test, with the filmstrip and my related commentary fresh in mind. (Preparatory to the laboratory, they would—or should—have read the appropriate sections of the laboratory manual.)

This time-tested procedure worked quite well in an academic setting, and I would judge it would also work in training young people in an in-house, professional setting.

Now, a few editorial observations about the Soiltest filmstrips. I believe, in general, the producers have let their zeal to sell equipment affect their directorial judgment. They cater to the very lowest of intelligence, I guess figuring that they'll have a broader market. In one frame, for example, they say, "This is a knife."

In some filmstrips, there are out-and-out mistakes, or at the least, poor advice. In the "Unconfined Compression Test" filmstrip, the narrator stumbles over a dial reading, and there is a strange and repetitious emphasis: "Be sure the strain dial is set to zero," an arbitrary and unnecessary requirement. Conversely, they do not mention the real need to be certain of *positive contact* of the strain apparatus at the initiation of load application.

Soiltest filmstrips can be a worthwhile investment, as useful teaching aids, if supplemented by appropriate review and commentary from experienced supervisors. A little humor can be helpful in softening the excesses of the commercialism. Soiltest has more than 20 filmstrips available, ranging from simple laboratory index testing techniques to the use of sophisticated instrumentation for seismic and resistivity field testing. For the more complex subject matter, such as the latter, they also offer geophysical training programs. Their corporate headquarters is Soiltest, Inc., 2205 Lee Street, Evanston, Illinois, 60202.

10.3.2 Encyclopaedia Britannica Filmstrips

Two of the finest sound filmstrips I have seen are from the Basic Earth Science Program, coproduced by Encyclopaedia Britannica and the American Geologic Institute, copyright 1969:

1. "Geologic Measurements and Maps," Series No. 6414.
2. "Investigating Rocks," Series No. 6415.

They are recommended without reservation and may truly be regarded as effective self-study materials. A young, conscientious person need merely learn the simple techniques for setting up the filmstrip in the projector, turn on the phonograph, and sit back and learn. These filmstrips (and probably many others) are available from: Encyclopaedia Britannica, 425 North Michigan Avenue, Chicago, Illinois, or 151 Bloor Street West, Toronto, Canada.

10.3.3 Miscellaneous References

1. *Soil Mechanics Laboratory: Procedures and Write-up,* E. J. Monahan, Ph.D., P.E. I have available an unpublished 90-page manual that I developed for the course described in the previous section. It is elementary and presumes no knowledge or experience in soils. People developing such courses at the growing number of community and county colleges, particularly BSET programs (Bachelor of Science in Engineering Technology), may find it helpful, since scheduling and a "To the Instructor" section is included, detailing (for example) an hour-by-hour description for the early (orientation) weeks of the course.

A variety of readers may be interested in the detailed treatment of procedures for soil identification and classification, an important topic that could not properly be described in detail in this book. Suggestions regarding report writing are also included. The manual is in printed form, but not of polished, published format.

2. *Basic Procedures For Soil Sampling and Core Drilling,* W. L. Acker III. This highly recommended document is distributed by the Acker Drill Co., P.O. Box 830, Scranton, Pennsylvania, 18501.

3. *The Design of Foundations for Buildings,* Johnson and Kavanaugh, McGraw-Hill, 1968. This reference is cited in my bibliography (No. 18), but for another reason. I mentioned in the text that detailed descriptions of load testing was "beyond the scope of this book." For the interested young field person, Johnson and Kavanaugh have an excellent treatment of this subject in Chapter 9.

4. *The Effects of Angularity on the Compaction and Shear Strength of a Cohesionless Material,* Richard Swiderski, Master's thesis, Newark College of Engineering, 1976. This master of science thesis on angularity and its effects on soil behavior is listed because I think the topic is important and has not been given the attention it deserves. Indeed, the results of the literature search conducted by its author, at my direction, confirmed my suspicion that very little had been done to investigate the subject. Hence, while the thesis does not purport to be a definitive, complete (or even "correct") treatment of the subject, it is truly "basic research." Another interesting feature of the investigation is that it was the combined effort of two excellent students, one undergraduate senior civil engineering student (Jeff Tubelo), and the author, a graduate student in civil engineering with an undergraduate degree in geology. The different but related undergraduate backgrounds contributed to an interesting and rewarding study.

10.4   Additional Seminars

I have available seminars in detail on the following:

1. *Sources of Free and Inexpensive Geotechnical Land-Use Planning and Design Information: Paper Reconnaissance.*
2. *Site Investigation Techniques:* (a) Field Reconnaissance, (b) Test Pit Inspection, (c) Field Testing, (d) Inspection, (e) Field Load Testing.
3. *Introductory Soils Laboratory Techniques.*
4. *Engineering-Geologic Field Trip: Tocks Island Site and Yards Creek Pumped Storage.*

References

1. C. B. Breed and G. L. Hosmer, *Elementary Surveying*, Vol. I, 8th ed., Wiley, New York, 1945.
2. V. J. Brown, "Soil Compaction in Narrow Places," Arrow Manufacturing Co., Denver, CO, 1967.
3. D. M. Burmister, *Soil Mechanics*, Vol. I, Columbia University Press, New York, 1955.
4. D. M. Burmister, "Suggested Methods of Test for Identification of Soils," *ASTM: Procedures for Testing Soils*, New York, 1958.
5. A. Brinton Carson, *General Excavation Methods*, Krieger, Huntington, NY, 1980.
6. H. R. Cedergren, *Seepage, Drainage, and Flow Nets*, 2nd ed., Wiley, New York, 1977.
7. Y. S. Chae and T. J. Gurdziel, "New Jersey Fly Ash As Structural Fill," *New Horizons in Construction Materials*, Vol. I, Envo Publ., Lehigh Valley, PA, 1976.
8. T. A. Coleman, "Polystyrene Foam Is Competitive, Lightweight Fill," *Civil Engineering*, February 1974, pp. 68–69.
9. A. Eggestad, "Experiences of Compaction Control in Sand and Gravel," *International Conference on Compaction*, Vol. II, Paris, 1980.
10. H. Fang (ed.), *New Horizons in Construction Materials*, Vol. I, Envo Publ., Lehigh Valley, PA, 1976.
11. G. A. Fletcher and V. A. Smoots, *Construction Guide for Soils and Foundations*, Wiley, New York, 1974.
12. T. R. Giddings, "A Rapid Method of Controlling Compaction By Plate Loading Tests," *International Conference on Compaction*, Vol. II, Paris, 1980.
13. D. M. Greer and D. C. Moorhouse, "Engineering—Geologic Studies for Sewer Projects," *Journal of the Sanitary Engineering Division, Proceedings, ASCE*, February 1968.
14. M. Healy, "Utilization of Marginal Lands," Master's project (unpublished), New Jersey Institute of Technology, Newark, NJ, 1975.
15. D. M. Hendron and L. Holish, "Quality Control of Earthwork Construction Using Cohesive Soils with Highly Variable Properties," *International Conference On Compaction*, Vol. II, Paris, 1980.
16. R. C. Hirschfeld, "Shear Strengths for Analysis of the Stability of Earth-dam Slopes," American Society of Civil Engineers Seminar, New York, June 2, 1965.
17. B. K. Hough, *Basic Soils Engineering*, 2nd ed., Ronald, New York, 1969.
18. S. M. Johnson and T. Kavanaugh, *The Design of Foundations for Buildings*, McGraw-Hill, New York, 1968.

19. J. Laguros and J. Robertson, "Compaction of Cover Material in Refuse Landfills," *International Conference on Compaction*, Vol. I, Paris, 1980.

20. T. W. Lambe, *Soil Testing for Engineers*, Wiley, New York, 1951.

21. T. W. Lambe, The Engineering Behavior of Compacted Clay, *Journal of the Soil Mechanics and Foundations Division, Proceedings, ASCE*, May, 1958.

22. W. W. Lowrance, *Of Acceptable Risk*, Kaufmann, Los Altos, CA, 1976.

23. W. C. McBee, D. Saylak, T. A. Sullivan, and R. W. Barnett, "Sulfur as a Partial Replacement for Asphalt in Bituminous Pavements," *New Horizons in Construction*, Vol. I, Envo Publ., Lehigh Valley, PA, 1976.

24. Maine State Highway Commission, "Insulation of Subgrade—Evaluation of First Year Data," Dow Chemical Co., Midland, MI, 1966.

25. A. Mardekian, W. Rowbotham, and B. Facente, "Soil Compaction Comparative Studies," Senior project (unpublished), New Jersey Institute of Technology, Newark, NJ, 1973.

26. R. E. Means and J. V. Parcher, *Physical Properties of Soils*, Merrill, Columbus, OH, 1963.

27. R. L. Meehan, "The Uselessness of Elephants in Compacting Fill," *Canadian Geotechnical Journal*, September 1967, pp. 358–360.

28. *Modern Plastics Encyclopedia*, McGraw-Hill, New York, 1968.

29. E. J. Monahan, Floating Foundation and Process Therefor, *U.S. Patent 3,626,702*, December 14, 1971.

30. E. J. Monahan, Novel Low Pressure Back-Fill and Process Therefor, *U.S. Patent 3,747,353*, July 24, 1973.

31. E. J. Monahan, "A Method for Specifying Percentage Soil Compaction," *Civil Engineering*, May 1974, pp. 68–69.

32. M. D. Morris, "Earth Compaction," *Construction Methods and Equipment*, McGraw-Hill Reprint, New York, 1959.

33. R. Peck, "Art and Science in Subsurface Engineering," *Geotechnique, International Journal of Soil Mechanics and Foundations*, March 1962.

34. R. Peck and H. Ireland, "Backfill Guide," American Society of Civil Engineers meeting, Jackson, MS, 1957.

35. R. B. Peck, W. Hanson, and T. Thornburn, *Foundation Engineering*, 1st ed., Wiley, New York, 1953.

36. R. R. Proctor, "Fundamental Principles of Soil Compaction," *Engineering News-Record*, August 31, September 9, September 21, September 28, 1933.

37. "Professional Liability Loss Control: Architects and Engineers," INAX Underwriters Agency Inc., Chicago, IL, 1980.

38. J. Reichert, "Various National Specifications on Control of Compaction," *International Conference on Compaction*, Vol. III, Paris, 1980.

39. F. Richards, *Sue the Bastards: Handbook for the Field Engineer*, Richards, Pittsford, NY, 1976.

40. L. J. Ritter and R. Paquette, *Highway Engineering*, 2nd ed., Ronald, New York, 1960.

41. W. L. Schroeder, *Soils in Construction*, 2nd ed., Wiley, New York, 1980.

42. G. F. Sowers, *Introduction to Soil Mechanics and Foundations: Geotechnical Engineering*, 3rd ed., Macmillan, New York, 1970.

43. G. F. Sowers, *Introduction to Soil Mechanics and Foundations: Geotechnical Engineering*, 4th ed., Macmillan, New York, 1979.

44. M. M. Truitt, *Soil Mechanics Technology*, Prentice-Hall, Englewood Cliffs, NJ, 1983.

45. U. S. Environmental Protection Agency, *Hazardous Waste Sites: National Priorities List*, HW-7.1, August 1983.

46. O. Wager and R. D. Holtz, "Reinforcing Embankments by Short Sheet Piles and Tie Rods," *New Horizons in Construction Materials,* Vol. 1, Envo Publ., Lehigh Valley, PA, 1976.

47. W. Williams, "Development and Use of Plastic Foam Insulation to Prevent Frost Action Damage to Highways—A Summary of Experience in the United States," *International Conference on Highway Insulation,* Wurzburg, Germany, May 1968.

48. E. J. Yoder, *Principles of Pavement Design,* Wiley, New York, 1959.

49. C. Zwingle, "On Specifying Percentage Compaction for Clay Soils," Master's project (unpublished), New Jersey Institute of Technology, Newark, NJ, 1981.

Index

AASHTO system, 9
Acceptable risks, 137
Acquisition, of borrow, 115
Aeolian, 50
Aeration, 96
Air-drying (laboratory compaction), 24
Air permeability, 89
Air photos, 116
Airport construction, 62
Allowable bearing pressure, 34
Alluvial, 11, 50
Aluminum (cans), 52
Ambiguity, avoidance, 146
Ames dials, reset, 150
Anchor bolts, 70
Angularity, 50, 90, 189
Anistropy, 48, 68, 80, 130
Approach fills, 61
Archimedes' principle, 75
Architect, 2, 115
Arrow hammer, 95
Artificial fills, 53
Artificial subbase, 58
Ash, fills, 52, 61
ASTM Compaction Requirements, 23
ASTM designations, 129
ASTM procedures, 67
Atterberg Limits, 8, 130
Augmentation techniques, 126

"Backfill Guide," 93
Bad weather, 101
Ballast, 89
Bamboo, 52
Base course, 122
Basement walls, 63, 91
Bay muds, 87

Bearing capacity, 120
    allowable, 3
    failure, 3, 84
    presumptive, 124
Bedrock, depth, 115
Bentonite, 47, 69
Blasting, 77, 117
Blister densities, 130
Blow counts, 5, 7
Blown-in-place foam fills, 54
Blunders, avoidance, 16, 43
Boiling, 72, 76
Boring(s), 115
Boring depths, 121
Boring inspection, 141
Boring program, 123
Borrow:
    area, mound, 103
    changing, 2
    heterogeneous, 103
    loading equipment, 116
    searching, 155
    sources, 116
    texture, changes, 105
Bound water, 48
Breaching, 73, 77
Bridging, in fills, 88
Broken stone, backfill, 58
Building codes, 78, 117
Burmister system, 9

California Bearing Ratio, 100
Capillarity, 71, 97
Capillary flow, 77
Capillary rise, 78
Case histories, 166
Catskill Acqueduct Case Study, 62

Caving, 73, 77
Certification, 114
  levels of, 139
Changing borrow, 26, 36, 39, 44, 88, 104, 111
Cheap jobs, 136
Chemical resistance, 65
Chimney drain, 82
Chunk samples, 130
Chunk-volume determinator, 133
Cinder fills, 53
Classification:
  plastic soils, 22
  of soils, 89
Classification system(s), 8
  auxiliary, 5
Clay(s), lean, fat, 21
Clay fill, 102, 106
Clay minerals, 51
Client, 114
Coefficient of permeability, 98
Cold weather, 106
Cold weather gear, 144
Cohesionless soils, 4
Cohesive soils, 4
Collapse, of clays, 93
Compacted clays, 68
Compacted density, condition, 22
Compaction:
  blunders, 13
  definition, 18
  field, 3
  granular soils, 110
  history, 13
  laboratory, 3
  modified Proctor, 15
  moisture content, 20, 26
  standard Proctor, 14
  standard and modified Proctor, 21
Compaction control, 111
  CBR, 111
  comprimeter, 111
  degree of compaction, 111
  field density testing, 111
  by index properties, 111
  plate loading, 111
  pocket penetrometer, 111
  proof rolling, 111
  nuclear method, 111
  special circumstances, 111
  water balloon, 111
Compaction curves, locations, shapes, 22
Compaction Data Book, 39, 47
  corrections, 49

Compaction equipment, 104, 117
Compactor(s), 88
  hand, 94
  types, 96
Communication with clients, 139
Compleat field man, 141
Completion time, 115
Compressibility, 48
Compressible clay, 60
Confined spaces, 66, 91, 107, 126
Consequences, distress, failure, 134
Consolidation, 60, 69, 71
Consolidation test, 130
Construction engineer, 115
Construction rubble, 52
Contact pressure, 120, 121
Contingency fee system, 178
Continuity principle, 76
Contract documents, 114
Contractor, 2, 30, 32, 114, 170
Core drilling, 189
Core wall, 71, 81
Coring operations, 161
Correction, 114
Cost:
  factors, 136
  pressures, 44
  real estate context, 57
  Styrofoam, 57
Critical exit hydraulic gradient, 75
Crust, New Jersey Meadows, 121
Cut-and-cover construction, 62
Cutoff wall, 71

D_{10} size, 11
Dams, 98
Darcy's law, 73, 98
Degree of compaction, 44
Dense sand, 93
Density corrections, 49
Density gradients, 67
Density-relative density test, 130
Density requirements, 116
Density test coordinates, 163
Density testing, how much?, 134
Development, research, 46
Dilatancy test, 11
Differential icing, of pavements, 64
Differential settlements, 119
Direction measurements, 146
Disclaimers, 140
Dispersal structure, 80
Distance measurements, 145
Drag, on piles, 60

Drainage characteristics, 115
Drainage designations, 4
Drilling mud, 51, 69
Driving forces, 85
Drum rollers, 89
Dry borrow, 106
Dry density, 20
Dry of optimum, 68, 79, 80
Drying, scarification, 117
Dumped backfills, 92
Durability, 55, 64
Dynamic compaction, 87, 90

Earth dam(s), 2, 71, 102
 caveat, 69
Earth structures, 2, 72
Earthquakes, 77
Earthwork contractors, 154
Earthwork equipment, 116
Effective weight, 75, 84
Elastic modulus, 108, 118
Elephants, 91
Elevation measurements, 145
Embankment, 80
 stability, 84
End-result specifications, 28, 100, 104, 116
End-result and suggested method, 104
Energy (of) compaction, 96
Energy control, 96
Energy sensitivity, 48
Engineering geologist, 83
Engineering properties, 3, 7, 109
 as design basis, 136
Engineering technologists, 114
Environmental factors, 134
Environmental Protection Agency, 87, 138
Epoxies, 70
Equipment imbalance, 153
Equipotential drop, 73
Equipotential lines, 73
Erosion, 79
Ethics, 1, 28, 30
Euclid truck(s), 91
 as compactors, 155
Excavated materials, 102
Excavated surfaces, 102
Excavating, 102
Excavation, ease of, 116
Excavation equipment, 116
Excess porewater pressure, 69
Exclusion, in laboratory compaction test, 24
Exculpatory clauses, 139
Exit hydraulic gradient, 75
Expansiveness, 48, 71

Expensive jobs, 136
Expert witness, 112, 178

Fat clays, 48
Favors, gratuities, 171
Floating foundation, 53
Flocculent structure, 80, 93
Flood control dams, 84
Flooding, 94
Floor loads, 60
Flotation, 55, 65
Flow lines, 73
Flow net, 73
Flow net element, 74
Flow net requirements, 73
Flow path, 74
Field clothing, 143
Field density checks, 105
Field density hole, 126
Field density techniques, 105
Field density tests, 15, 126, 159
 locations, 105
Field engineer, 113
Field equipment lists, 142
Field moisture, 100
 inspection, 125
Field notes, 146
Field observations, 162
Field performance, 105
Field sampler, 67
Field testing costs, 134
Field trip, 190
Fill, 5
 condition, 6, 86, 102
 controlled, 22, 86
 correction measures, 107
 load-bearing, 16, 17, 86
 partially controlled, 22, 86
 placement, 103
 special, 86
 texture, 6, 86, 93, 102
 uncontrolled, 22, 86
 undulation, 18, 26
 use of, 102
Fill compaction, 93
Fill control procedures, 119
Fill inspector, 2, 43
Fill records, 163
Filmstrips, 187
Fines, 8, 102, 111
Foam plastic:
 fills, 53
 poured, 182
 rigid, 182

Foundation:
  design, 123
  failure, 112
  selection, 123
  suitability, 118
Freeze-thaw tests, 57
Frozen soil, 106
Front-end loader, 92
Frost action, 11, 48, 72, 77
Frost problems, 62
Frost-susceptible soils, 78

Garbage, 52
Garbage In, Garbage Out, 141
Geodex, 91
Geologic land forms, 118
Geologic shortcomings, 57
Geologists, 160
Geomorphology, 118
Geostick, 67, 124
Geotechnical Division (ASCE), 135
Geotechnical engineer, 2
Glacial drift, 62
Glacial till, 117
Glass, as fill, 52
Gradation, 50
Grade beams, 70
Graded filter, 12, 79, 82
Grade-separation case study, 61
Grain-size limitations, 102
Granular, 11, 92
Grout, 70
Grout curtain, 97

Hand tamper, 125
Hand sieving, 148
Harvard compaction, 47
Haul distance, 116
Hauling equipment, 116
Hazardous fills, 138
Hazen's effective size, 49
Head loss, 73
Headwater, 73
Heel test, 125
Highway(s), 58
  insulation, 56
  pavement systems, 56
Honeycomb structure, 80, 93
Horizontality measurements, 147
Hydraulic fills, 87, 90
Hydraulic gradient, 75
Hydrostatic pressure, 69
Hypothetical case histories, 57

Ice lenses, 77
Impact, 93
Imperviousness, 71
Index properties, 3, 6, 109
  as design basis, 136
Index property variations, 41, 46
  angularity, 40
  Hazen's effective size, 40
  percentage fines, 40
  plasticity index (PI), 40
  uniformity coefficient, 40
Industry-university interaction, 135
Infrastructure repair, 78
Injected backfill, 59, 66
In situ, 45
Inspection, 119
Inspector, 113, 114, 125, 128, 138, 170, 171
Instant moisture balance, use, 25
Insulation, 60

Jacked sample method, 130
Jetting, 94
Joints, faults, 161

Kaolinite, 47

Laboratory procedures, 189
Laboratory techniques, 190
Labor costs, 92
Laminar flow, 73, 98
Laws and regulations, 117
Lawsuits, 112, 139
Leachate control, 87
Lean clays, 48
Legal consequences, 44
Legal costs, 177
Length-diameter ratio, 68
Level adjustments, 149
Liability insurance, 177
Lift preparation, 103
Lift thickness, 88, 96, 100, 103, 104
Lightweight fills, 53, 61
Lime-stabilized soils, 107
Limitations, reasonable, 140
Liquefaction, 48, 72, 77
Liquid limit, 68, 111
Litigious society, 112
Live loads, 59
Load-bearing fills, 101
Loaded area shapes, 120
Load-settlement curves, 125
Load testing, 141, 148, 189
  setup, 149
Long term compression, 65

Loose sand, 93, 109
Love Canal, 138
Lunch hour filling, 107

Making do, 91
Major problems, 28
    ninety-five percent fixation, 31
    standard-modified ignorance, 28
Marble dust, 47
Marginally permeable, 72
Mechanical compactors, 66
Methods specifications, 28, 100
Mineralogy, 102
    differences, 109
Modified Proctor, 66
Moisture and energy sensitivity, 23, 26
Moisture computation, 96
Moisture content, 104
Moisture control, 96
    adjustments, 103, 107
Moisture sensitivity, 48, 90
Montmorillonite, 69

National Dam Inspection Program, 72
Negative skin friction, 60
Non-load bearing fills, 101
Nonplastic, fines, 47
Nonspecialists, 1, 2, 31, 32, 99
Nuclear density method, 133
Nuclear regulation, 134

Observer, 113
Odors, 87
Open gravel, 98
Optimum moisture content, 14, 22, 26
    corrections, 49
Oral records, 164
Organics, 88
OSHA, 118, 138
Ottawa sand, 110, 128
Outlet blanket, 82
Overturning, 84

Pace length, 145
Paper reconnaissance, 190
Passes, number of, 100, 104
Pavement system, 108
Payment, terms, 115
Penstock, 98
Percentage compaction, 100, 104
    for clays, 68
    example, 39
    method of determining, 34
Percent fines, 47

Percent saturation, 26
Permanency, 55, 64
Permeability, 49
Personal safety, 172
Photo records, 159, 164
Phreatic Surface, 79, 82
PI, of fines, 47
Pickford Bridge, 54, 56, 65, 182
Piezometer, 73, 97
Pile bearing capacity, 61
Pile driving, 77
Pile load, test, 60
Pipelines, 59
Piping, 11, 73, 76, 81, 84
Placement equipment, 116
Plastic, fines, 48
Plasticity, 11
Plasticity index, 102
Plate bearing tests, 108, 125
Plate loading test, field, 109
Plumb measurements, 147
Pneumatic rollers, 89
Pocket penetrometer, 124
Pole installations, 56
Poleset, 56, 95, 181
Polystyrene, 55
Polyurethane foam, 61
Porewater pressures, excess, 48, 84
Possibility context, 57
Postcompaction precautions, 106
Postconstruction damage, 139
Precast foam fills, 54
Precompaction equipment, 116
Preconsolidation, 54, 70, 124, 177
Pressure bulb, 120
    footing, 122
    mat foundation, 122
    overlapping, 123
    piles, 123
    standard, 120
    spiked heel, 122
    tire, 122
    uses, exceptions, limitations, 124
Pressure head, 73
Presumptive bearing capacity, 67
Proctor density, 107
Professional liability insurance, 112
Proof rolling, 118
Puddling, 90, 94
Pumped-storage, 84, 98
Pumping, 77

Qualitative fill control, 165
Quantity, of borrow, 115

Quartering, soil samples, 26
Quartzitic minerals, 51
Quick condition, 72
Quicksand, 76

Rain problems, 106
Random selection technique, 165
Rapid drawdown, 84
Rapid loading, clay, 124
Recompaction effects, 24
Reconnaissance, 141
Record keeping, 153, 162
Reinforced earth, 63
Relative density, 35, 82, 92, 100
Report writing, 162, 189
Representative batches, preparation, 24
Representative samples, 103, 158
Research, 46
Resisting forces, 85
Retaining structures, 62, 91, 93
Richter scale, 12
Road bases, 100, 107
Roadway embankments, 101
Rock, 82
    angle of placement, 83
    angularity, 82
    blocky, 82
    estimate, quantity, 116
    fissures, 98
    geologic defects, 83
    geologic soundness, 82
    interlocking, 82
    mineralogy, 83
    point contacts, 83
    rippability chart, 117, 173
    ripping, blasting, 88, 117, 173
    shape, 82
    slabby, 82, 88
    structure, 83
    surface texture, 82
    weathering, 83
    wetting, 83
Roofing, 73, 77
Route construction, 58, 78
Rubber, as fill, 52
Rubble fills, 88

Sample bags, 147
Sample disturbance, 133
Sampling ratio, 134
Sand cone apparatus, 154
Sand cone method, 126
Sand cone test, steps, 128
Sand drains, 54, 70, 177

Sanitary fills, 87
Scarification, drying, 96, 103
Scraper, 117, 154
Sediments, 62
Seepage, 81
    losses, 71, 75
    pressures, 48, 74, 75
    quantity, 74
Select fill, 10, 21, 23, 26, 49, 80, 81, 90, 102
Seminars available, 190
Settlement(s), 71, 109
    computation, 92
    plates, 105, 118
    unacceptable, 3, 110
Shape factor, 74
Shear failure, 84
Shear strength, 68
Sheepsfoot roller, 83, 89
Sheet pile, 97
Shell, of dam, 71
Shocks, 77, 93
Sieve numbers, 69
Sieve tests, 130
Silts, 22, 83
    frost action, 84
    liquefaction, 84
    moisture sensitivity, 84
    seepage pressures, 84
Simulation, of field conditions, 96, 127
Sinkholes, 160
Site investigation, 190
Slab-on-grade, 70
Sliding, 84
Slope stability, 63, 79, 84
Slurry, 69
Specifications, 3, 99
    authority, 112, 113
    careless, 33
    enforcement, 111, 172
    evaluation, 115
    implementation, 111
    nontechnical aspects, 114
    responsibility, 112, 113
    special criteria, 107
Soak tests, 57
Soft wet clay, 93
Soil classification, visual, 148
Soil Conservation Service (SCS), 115
Soil identification, classification, 189
Soil pins, 65
Soil maps, 115
Soil sampling, 189
Soil stabilization, 54
Specialist's services, 116

Specific gravity, 20, 51
Speedy moisture device, 128
Splitting, soil samples, 27
Sprinkling, field, 96, 117, 155
Stability, 6, 71
   of excavations, 167
Stabilization, 57
Standard Penetration Test, 34, 45
Static loadings, 109
Stream terrace, 82
Strength-deformation tests, 130
Strip mines, 157
Stripping and grubbing, 29, 102, 116
Structural beams, 70
Structural engineer, 2, 99, 119, 122
Structural fill, 94, 101, 102
Structural performance, 56
Structural slab, 60, 70
Structure of clays, 79
Styrofoam HI, 54
Subbases, 107
Subgrade, 55, 81, 100, 122
Subgrade modulus, 108
Subgrade stabilization, 58
Submerged weight, 75
Subsidence, 160
Subsurface conditions, 44
Suggested guide specifications, 101
Suggested-method and end result, 100
Sulfur, 52
Sumps, 87
Supercompactors, 90, 125
Surcharge, 70
Surface damage, 106
Surveying techniques, 146
Swamp muck, 53

Taccoa Dam, 72, 77
Tailwater, 73
TALB, 21
Tamping rollers, 90
Tare weight, 175
Target value density, 28, 100, 103, 126, 159
Techonomics, 135, 176
Tectonic forces, 175
Terzaghi, 110
Teton Dam, 72, 77
Textural tests, 130
Texture, 22, 111, 118
   acceptable, 102
   of borrow, 115
   conformance, 127
Time, pressures, 44
Times Beach, 138

Torvane, 124
Total cost context, 57
Test pits, 115
   inspection, 86, 141
   locations, 165
   sampling, 141, 147
Test strips, 104
Traditional practice (causing problems), 42
   architects and owners, 43
   construction engineers, contractors, 42
   nonspecialists, their roles, 42
   structural engineer, 42
Trafficking problems, 116
Traffic vibrations, 93
Tremie, 45, 91
Trench backfill, 92, 95
Triaxial compression test, 130

Unacceptable end results, 114
Unconfined compressive strength, 5, 11, 68, 111, 124, 130
Unconsolidated alluvium, 62
Undulation of fill, 105
Unified Soil Classification System, 8, 50, 90
Uniformity coefficient, 45, 50
Uplift pressure, 69
Unsuitable material, 102
Utility installations, 56

Vaiout Reservoir disaster, 137
Varved clays, 121, 175
Vector control, 87
Vegetation cover, 87
Velocity head, 73
Venting, gases, 87
Verification, degree of, 134
Very large jobs, 184
   management, 185
   quality control, 185
   statistical methods, 186
Vibrations, 109
   postconstruction, 110
Vibratory rollers, 90
Vibroflotation, 87, 90
Void ratio, 45, 92
Volcanic clays, 53
Volumeasure, 129

Washington Densometer, 129
Waste materials, 52
Water absorption, 55, 66
Water balloon method, 129
Water table, depth, 115

Weather, 44
Weather restrictions, 105
Weep holes, 94
Well points, 87
Weight credit, 53, 60, 61, 95
Weight-credit update, 181

Westway, 184
Wet borrow, 105
Wet of optimum, 68, 79, 80
Wicks, 70, 177
Work changes, 115
Written criticism, 166